Frontiers in Mathematics

Friedrich Kasch
Adolf Mader

Rings, Modules, and the Total

Birkhäuser Verlag
Basel • Boston • Berlin

Authors' addresses:

Friedrich Kasch
Mathematisches Institut
Universität München
Theresienstr. 39
80333 München
Germany
e-mail: Friedrich.Kasch@t-online.de

Adolf Mader
Department of Mathematics
University of Hawaii
2565 The Mall
Honolulu, HI 96822
USA
e-mail: adolf@math.hawaii.edu

2000 Mathematical Subject Classification 16D10

A CIP catalogue record for this book is available from the
Library of Congress, Washington D.C., USA

Bibliographic information published by Die Deutsche Bibliothek
Die Deutsche Bibliothek lists this publication in the Deutsche National-
bibliografie; detailed bibliographic data is available in the Internet at
<http://dnb.ddb.de>.

ISBN 3-7643-7125-0 Birkhäuser Verlag, Basel – Boston – Berlin

Part of Springer Science+Business Media
Cover design: Birgit Blohmann, Zürich, Switzerland
Printed on acid-free paper produced from chlorine-free pulp. TCF ∞
Printed in Germany
ISBN 3-7643-7125-0

9 8 7 6 5 4 3 2 1

www.birkhauser.ch

Contents

Preface

This book is an introductory text to the theory of rings and modules with special emphasis on decomposition properties. The total is a concept that is as fundamental as that of the radical, and has not appeared in text books before. Except for the most basic concepts and facts of ring and module theory, the book is self–contained, and proofs are included even of rather well–known results. Yet, the book reaches into the frontier of the subject.

What are the best elements in a ring R (always with 1)? Obviously, these are the invertible elements that are the divisors of 1. We call the divisors of an idempotent $0 \neq e = e^2$ *partially invertible*, so r is partially invertible if there exists $s \in R$ such that $e = sr$. In this book we intend to show that the partially invertible elements are the "second best" elements in a ring. The set of elements of R that are not partially invertible we denote by $\mathrm{Tot}(R)$ and call it the *total* of R.

These notions are not restricted to rings. If Mod–R is the category of all unitary right R–modules $A, B, C, M, W, \ldots$, then $f \in \mathrm{Hom}_R(M, W)$ is called partially invertible if there exists $g \in \mathrm{Hom}_R(W, M)$ such that

$$\mathrm{End}(M) \ni e := gf = e^2 \neq 0,$$

which again means that f is a divisor of a non–zero idempotent. This is equivalent to the fact that there exist non–zero direct summands $A \subseteq^{\oplus} M$, $B \subseteq^{\oplus} W$ such that f induces an isomorphism

$$A \ni a \mapsto f(a) \in B.$$

Homomorphisms $f \in \mathrm{Hom}_R(M, W)$ that are not partially invertible are called *total non–isomorphisms*. The set of total non–isomorphisms on M to W is called the *total on M to W* and denoted by $\mathrm{Tot}(M, W)$. These notions were defined by the first author of this book in 1982 and studied in his seminar in Munich. They generalize special concepts used in the theory of modules with local endomorphism rings. The first publication on the total in this generality was a PhD thesis by W. Schneider ([27]). In the following years several authors dealt with the concepts or used them (Beidar [4], Beidar and Kasch [5], [6], Beidar and Wiegand [7], Kasch [16], [17], [18], [19], [20], Kasch and Schneider [21], [22], [27], Zelmanowitz [28], Zöllner [29]). By now there are numerous results and most of them are covered in this book.

In Chapter II we derive the fundamental properties. The total is in general not additively closed but it has the following multiplicative closure property:

If $X, Y \in \text{Mod--}R$ *and* $h \in \text{Hom}_R(X, M)$, $k \in \text{Hom}_R(W, Y)$, *then* $k\,\text{Tot}(M, W)h \subseteq \text{Tot}(X, Y)$.

A set of maps with this property is called a *semi–ideal* in Mod–R. Thus, for a ring R,

$$x, y \in R \Rightarrow x\,\text{Tot}(R)y \subseteq \text{Tot}(R).$$

Although $\text{Tot}(M, W)$ is not additively closed, it has certain additive closure properties. We introduce (Definition 2.2)

$$\text{Rad}(M, W) := \{f \in \text{Hom}_R(M, W) \mid \forall\, g \in \text{Hom}_R(W, M), fg \in \text{Rad}(S)\}.$$

The radical $\text{Rad}(M, W)$ contains the usual radicals of $\text{Hom}_R(M.W)$ both as a left $\text{End}_R(M)$–module and as a right $\text{End}_R(W)$–module. Also, for a ring R identified naturally with $\text{Hom}_R(R_R, R_R)$ we have $\text{Rad}(R_R, R_R) = \text{Rad}(R)$. It is now true that

$$\text{Rad}(M, W) + \text{Tot}(M, W) = \text{Tot}(M, W),$$

which also implies that $\text{Rad}(M, W) \subseteq \text{Tot}(M, W)$. In particular, in the ring case, $\text{Rad}(R) \subseteq \text{Tot}(R)$.

Not only is it true that $\text{Rad}(M, W) \subseteq \text{Tot}(M, W)$, but also the *singular submodule* $\Delta(M, W)$ (the set of all $f \in \text{Hom}_R(M, W)$ with large kernel) and the *co–singular submodule* $\nabla(M, W)$ (the set of all $f \in \text{Hom}_R(M, W)$ with small image) are contained in $\text{Tot}(M, W)$. A natural question arises: For which M and W is it true that $\text{Rad}(M, W) = \text{Tot}(M, W)$ or $\Delta(M, W) = \text{Tot}(M, W)$ or $\nabla(M, W) = \text{Tot}(M, W)$? Since Rad, Δ, and ∇ are additively closed, in the case that any one of the above equalities holds, the total is also additively closed. For the question of equality we provide some interesting answers in Chapter III (see III.Theorem 2.2, III.Corollary 2.5). Also exchange properties for modules imply good additive properties for the total.

In Chapter IV we present a difficult part of algebra. If a module M has a decomposition

$$M = \bigoplus_{i \in I} M_i$$

where all endomorphism rings $S_i := \text{End}(M_i)$, $i \in I$, are local rings, then the decomposition is called an *LE–decomposition.*

Modules with LE–decompositions were studied by many algebraists and success in the subject required hard work. By now there exists a satisfactory theory of these modules. We present the main part of this theory. If $S := \text{End}(M)$, then the existence of an LE–decomposition implies that $\text{Tot}(S)$ is an ideal in S, hence $S/\,\text{Tot}(S)$ is again a ring.

First Main Theorem (IV.3.3). *Assume that* M *is a module with an LE–decomposition, and set* $S := \text{End}(M)$. *Then the quotient ring* $S/\,\text{Tot}(S)$ *is isomorphic to a product of endomorphism rings of vector spaces over division rings.*

Examples (IV.Example 2.5) show that for LE–decompositions Rad(S) need not be equal to Tot(S). Hence for the First Main Theorem, the total is truly the essential notion. But if Rad(S) = Tot(S), then there are further "very good" properties.

Second Main Theorem (IV.4.1). *Assume that M is a module that has an LE–decomposition, and set $S := \mathrm{End}(M)$. Then the following statements are equivalent.*

1) $\mathrm{Rad}(S) = \mathrm{Tot}(S)$.
2) *Every LE–decomposition of M complements direct summands.*
3) *Any two LE–decompositions of M have the replacement property.*
4) *If $M = \bigoplus_{i \in I} M_i$ is an LE–decomposition, then the family $\{M_i \mid i \in I\}$ is a local semi–t–nilpotent family.*

A direct decomposition $M = \bigoplus_{i \in I} M_i$ *complements direct summands* if for every direct summand A of M there is a set of indices $J \subseteq I$ such that

$$M = A \oplus \bigoplus_{j \in J} M_j.$$

Two LE–decompositions

$$M = \bigoplus_{i \in I} M_i = \bigoplus_{j \in J} N_j$$

satisfy the *replacement property* if *for each subset $I_0 \subseteq I$ there exists a subset $J_0 \subseteq J$ such that*

$$M = \left(\bigoplus_{i \in I \setminus I_0} M_i\right) \oplus \left(\bigoplus_{j \in J_0} N_j\right).$$

A family of LE–modules (= modules with local endomorphism ring) $\{M_i \mid i \in I\}$ is called *locally semi–t–nilpotent* if and only if for every infinite sequence of pairwise different elements

$$i_1, i_2, i_3, \ldots \in I$$

and for every sequence of homomorphisms

$$f_1, f_2, f_3, \ldots \text{ with } f_j \in \mathrm{Tot}(M_{i_j}, M_{i_{j+1}})$$

and for every $x \in M_{i_1}$, there exists $n \in \mathbb{N}$ such that

$$f_n f_{n-1} f_{n-2} \cdots f_2 f_1(x) = 0.$$

Who was involved in formulating and proving this theorem? It is difficult to give a precise answer including all progress and names. We believe that Harada developed a major part of the theory of LE–decompositions. Therefore we only cite one paper by Harada ([12]) that will serve as a source of further information. After Harada (1974) proofs were improved by several authors (e.g. [29]).

In the last chapter we consider completely decomposable Abelian groups that constitute the simplest interesting class of torsion–free Abelian groups. After providing the necessary background for the novice in this subject, we compute the total of the endomorphism ring in a special case (V.Theorem 2.1) and describe a recursive method for determining which maps in the endomorphism ring of a completely decomposable group belongs to the total.

In a final section we discuss torsion–free Abelian groups of finite rank in general. These groups have very ill–behaved direct decompositions (V.Example 3.1) but by moving to an associated category one obtains LE–decompositions and the total of an endomorphism ring in this category is just the radical (V.Corollary 3.16, V.Theorem 3.17).

We hope that this monograph will serve as an introduction to the total and form the basis for further progress.

Chapter I

General Background

We consider rings R always with identity $1 = 1_R$. The category of right R–modules is denoted by Mod–R. Maps are written on the left.

If $f \in \mathrm{Hom}_R(M, W)$, then $\mathrm{Im}(f)$ and $\mathrm{Ker}(f)$ denote, respectively, the image and the kernel of f. The following theorem collects the Isomorphism Theorems that are a fundamental tool in module theory and will be used without explicit mention.

Theorem 0.1. *Let R be a ring and let A, B, C be R–modules.*

1) *Let $f \in \mathrm{Hom}_R(A, B)$. Then $\mathrm{Ker}(f) = \{a \in A \mid f(a) = 0\}$ is a submodule of A and*

$$A/\mathrm{Ker}(f) \cong f(A).$$

2) *Let A, B be submodules of an R–module M. Then*

$$(A + B)/B \cong A/(A \cap B).$$

3) *Let $A \subseteq B \subseteq C$ be a chain of R–modules. Then*

$$(C/A)/(B/A) \cong C/B.$$

4) **The Correspondence Theorem.** *Let $A \subseteq B \subseteq M$ be a chain of R–modules. Then there is a bijection*

$$\{C \mid A \subseteq C \subseteq B\} \ni C \mapsto C/A \in \{D \mid D \subseteq B/A\}.$$

The emphasis in this book is on direct decompositions of modules. Suppose that M is an R–module and

$$M = K \oplus L. \tag{1}$$

Associated with the decomposition (1) are several maps, namely the standard **projections** $\pi_K : M \to K$ **along** L and $\pi_L : M \to L$ along K. Let ι_K and ι_L denote the inclusion maps of K and L in M, respectively. We distinguish the projections from the maps $e_K = \iota_K \pi_K \in \mathrm{End}_R(M)$ and $e_L = \iota_L \pi_L \in \mathrm{End}_R(M)$ that we call the **projectors** belonging to (1). These are idempotent endomorphisms of M. They are **orthogonal** meaning that $e_K e_L = 0$ and $e_L e_K = 0$, and they form a **complete system of idempotents** in the sense that $e_K + e_L = 1_M$ where 1_M denotes the identity map on M.

Conversely, suppose that $e \in \mathrm{End}_R(M)$ is an idempotent, i.e., $e^2 = e$. Then $1-e$ is an idempotent in $\mathrm{End}_R(M)$ and $\{e, 1-e\}$ is a complete system of orthogonal idempotents. Furthermore,

$$M = e(M) \oplus (1-e)(M).$$

We will need in Chapter IV an extension to this standard fact.

Lemma 0.2. *Given homomorphisms*

$$f : A \to M, \quad g : M \to B,$$

such that gf is a non-zero isomorphism, then

$$M = \mathrm{Im}(f) \oplus \mathrm{Ker}(g). \tag{2}$$

If M is indecomposable, then f and g are also isomorphisms. Equivalently, if M is indecomposable and f is not an isomorphism, then gf is not an isomorphism.

Proof. Since gf is an isomorphism, f is injective and g is surjective. Let $m \in M$. Then, gf being surjective, there is $a \in A$ such that $g(m) = gf(a)$. It follows that $m - f(a) \in \mathrm{Ker}(g)$ hence $m = f(a) + (m - f(a)) \in \mathrm{Im}(f) + \mathrm{Ker}(g)$, i.e., $M = \mathrm{Im}(f) + \mathrm{Ker}(g)$. Assume that $f(a) \in \mathrm{Im}(f) \cap \mathrm{Ker}(g)$. Then $gf(a) = 0$, hence $a = 0$ and $f(a) = 0$. This means that $\mathrm{Im}(f) \cap \mathrm{Ker}(g) = 0$.

If M is indecomposable, then $\mathrm{Ker}(g) = 0$ and $\mathrm{Im}(f) = M$, so g is injective in addition to being surjective and f is surjective in addition to being injective, i.e., they are both isomorphisms. □

In a special case one can intersect a direct decomposition with a submodule and obtain a direct decomposition of the submodule. A submodule K of the module M is **fully invariant** in M if $f(K) \subseteq K$ for every endomorphism f of M.

Lemma 0.3. *Let $M = A \oplus B$ be a direct decomposition of R-modules, and let K be a fully invariant submodule of M. Then $K = K \cap A \oplus K \cap B$.*

Proof. Let e be the idempotent in $\mathrm{End}_R(M)$ with the property that $e(M) = A$ and $(1-e)(M) = B$. Then $e(K) \subseteq K \cap A$ and $(1-e)(K) \subseteq K \cap B$. Let $x \in K$. Then $x = e(x) + (1-e)(x) \in K \cap A + K \cap B$. Hence $K = K \cap A + K \cap B$ and the sum is obviously direct. □

If A is a direct summand of M, then there may or may not be several direct complements of A in M. Given

$$\text{(I)} \qquad M = A \oplus B$$

we describe the possible modules C with

$$\text{(II)} \qquad M = A \oplus C$$

by using the elements of $\operatorname{Hom}_R(B, A)$. In fact there is a bijective mapping between $\operatorname{Hom}_R(B, A)$ and the set of all possible C in (II). First, using (II), there is a unique homomorphism $\varphi_C : B \to A$ such that $C := (\varphi_C + 1)(B) := \{\varphi_C(b) + b \mid b \in B\}$. Conversely, for every $\varphi \in \operatorname{Hom}_R(B, A)$, the set

$$\text{(III)} \qquad B_\varphi := (\varphi + 1)(B) := \{\varphi(b) + b \mid b \in B\}$$

is a direct complement of A in M: $M = A \oplus B_\varphi$. The following is a generalized version of [24, Lemma 2.5].

Lemma 0.4. *Let $M = A \oplus B$ be a direct decomposition of R–modules. Denote by $\mathcal{C}$ the set of all direct complements of A in M. Then*

$$\operatorname{Hom}(B, A) \ni \varphi \mapsto B_\varphi := \{\varphi(b) + b : b \in B\} \in \mathcal{C}$$

defines a bijective mapping and

$$\varphi + 1 : B \ni b \mapsto \varphi(b) + b \in B_\varphi$$

is an isomorphism.

Proof. Let $M = A \oplus B$ and $\varphi \in \operatorname{Hom}_R(B, A)$. We first show that $\varphi \mapsto (\varphi+1)(B) := \{\varphi(b) + b \mid b \in B\}$ is a well–defined map assigning $\varphi \in \operatorname{Hom}_R(B, A)$ a complement of A in M. It is routine to check that $\varphi + 1 : B \ni b \mapsto \varphi(b) + b \in (\varphi + 1)(B)$ is an isomorphism, and that $A \cap (\varphi + 1)(B) = 0$. Finally, if $x \in M$, then $x = a + b$ for some $a \in A$ and $b \in B$, and hence $x = (a - \varphi(b)) + (\varphi + 1)(b) \in A + (\varphi + 1)(B)$. So $M = A \oplus (\varphi + 1)(B)$.

We show next that every complement is obtained in this fashion. Suppose that $M = A \oplus B = A \oplus C$. Let $\pi : M \to A$ be the projection of M onto A with kernel C. Then

$$\forall x \in M,\ x = \pi(x) + (x - \pi(x)), \quad \pi(x) \in A, x - \pi(x) \in C. \tag{3}$$

Let $\phi_C := -\pi \restriction_B$. Then $\phi_C \in \operatorname{Hom}_R(B, A)$. By (3) $(\phi_C + 1)(B) \subseteq C$. Let $c \in C$ and write $c = a + b$ with $a \in A$ and $b \in B$. Then $b = -a + c$ uniquely, so $\phi_C(b) = a$ and $c = \phi_C(b) + b \in (\phi_C + 1)(B)$. This shows that $C = (\phi_C + 1)(B)$. As before $\phi_C + 1 : B \ni b \mapsto \phi_C(b) + b \in C$ is an isomorphism. $\square$

More generally we deal with arbitrary direct sums

$$M = \bigoplus_{i \in I} M_i. \tag{4}$$

A family $\{M_i \mid i \in I\}$ of submodules of M is an **independent family** if they generate a direct sum $\sum_{i \in I} M_i = \bigoplus_{i \in I} M_i$ (not necessarily equal to M). An independent family $\{M_i \mid i \in I\}$ of submodules of M **finitely generates direct summands** if for every finite set $J \subset I$, the sub–sum $\bigoplus_{j \in J} M_j$ is a direct summand of M. For example, $\bigoplus_{i \in I} M_i \subseteq \prod_{i \in I} M_i =: M$ finitely generates direct summands of M.

The direct sums thus far discussed were **internal direct sums**, i.e., we considered submodules M_i of a given module so that $\sum_i M_i$ made sense and the question left was whether the sum was direct. It is also possible to start with an arbitrary family M_i of R–modules, form the Cartesian product of the M_i and define addition and scalar multiplication component–wise to obtain the **direct product** $\prod_i M_i$ of the modules M_i. The subgroup of all finitely non–zero vectors of $\prod_i M_i$ is the **external direct sum** denoted by $M = \dot{\bigoplus}_i M_i$.

We frequently use the Dedekind Identity or Modular Law: If N, M, K are submodules of some module U and $N \subseteq M$, then

$$M \cap (N + K) = N + (M \cap K). \tag{5}$$

A useful and attractive version of the Dedekind Identity is the following short exact sequence with natural maps

$$0 \to \frac{M \cap K}{N \cap K} \to \frac{M}{N} \to \frac{M + K}{N + K} \to 0 \tag{6}$$

where M, N, K are submodules of some module U and $N \leq M$.

When we deal with modules or other algebraic structures, $A \subseteq B$ normally means that A is a substructure of B, not just a subset. The exceptions will be clear from the context. The symbol $A \subseteq^{\oplus} M$ means that A is a direct summand of M. A module S is **small** or superfluous in the module M if $S + A = M$ implies that $A = M$. We write $S \subseteq^{\circ} M$ in this case. Note that $S \subseteq^{\circ} M$ and $f \in \mathrm{Hom}_R(M, W)$ implies that $f(S) \subseteq^{\circ} W$. A module $L \subseteq M$ is **large** or essential in the module M if $L \cap A = 0$ implies that $A = 0$. We write $S \subseteq^{*} M$ in this case.

An easy property of large submodules will be needed later.

Lemma 0.5. *Let A, B, C be submodules of some module M. If $A \cap B = 0$ and $B \subseteq^{*} C$, then $A \cap C = 0$.*

Proof. Suppose that $A \cap C \neq 0$. Then, B being large in C, $0 \neq B \cap (A \cap C) \subseteq A \cap B = 0$, a contradiction. □

A module M is **simple** if it has no submodules other than 0 and M. A module M is **semisimple** if it is the direct sum of simple submodules.

Theorem 0.6. *Let* $M \in \text{Mod–}R$*. Then the following statements are equivalent* ([15, 8.1.3]).

1) *M is semisimple.*
2) *Every submodule of M is a sum of simple submodules.*
3) *M is a sum of simple submodules.*
4) *Every submodule of M is a direct summand of M.*
5) *M contains no proper large submodules.*

A ring R is **semisimple** if R_R is semisimple.

Theorem 0.7. *Let R be a ring. The following statements hold* ([15, 8.2.1, 8.2.2]).

1) *R is semisimple if and only if ${}_RR$ is semisimple.*
2) *If R is semisimple, then every R–module is semisimple.*

Let $M \in \text{Mod–}R$. The **radical** $\text{Rad}(M)$ is defined by

$$\text{Rad}(M) := \bigcap\{A \mid A \text{ is maximal in } M\}.$$

Important properties of the radical are listed as a theorem.

Theorem 0.8. ([15, Theorem 9.1.1, Theorem 9.1.4, Theorem 9.4.1])

1) $\text{Rad}(M) = \sum\{B \mid B \subseteq^\circ M\}$.
2) $\text{Rad}(M/\text{Rad}(M)) = 0$.
3) *If M is finitely generated, then* $\text{Rad}(M) \subseteq^\circ M$.

The analogous statements are valid for left R–modules. The radical of a ring R is defined by $\text{Rad}(R) := \text{Rad}(R_R)$. The relevant properties of the radical of a ring are collected in another theorem.

Theorem 0.9. ([15, Theorem 9.3.2, Lemma 9.3.1, Theorem 9.2.1] *The following statements hold.*

1) $\text{Rad}(R) = \text{Rad}({}_RR)$ *and* $\text{Rad}(R)$ *is a two–sided ideal of R.*
2) *Let A be a right ideal of the ring R. $A \subseteq \text{Rad}(R)$ if and only if $\forall\, a \in A$, $(1-a)^{-1} \in R$.*
3) *For $M \in \text{Mod–}R$, $M\,\text{Rad}(R) \subseteq \text{Rad}(M)$.*

The dual notion to the radical is the socle.
Let $M \in \text{Mod–}R$. The **socle** $\text{Soc}(M)$ is defined by

$$\text{Soc}(M) := \sum\{A \mid A \text{ is minimal } (= \text{simple}) \text{ in } M\}.$$

Basic properties of the socle are listed as a theorem.

Theorem 0.10. ([15, Theorem 9.1.1, Theorem 9.1.3, Theorem 9.1.4])

1) $\mathrm{Soc}(M) = \bigcap\{B \mid B \subseteq^* M\}$.
2) $\mathrm{Soc}(M)$ *is the largest semisimple submodule of* M.
3) $\mathrm{Soc}(\mathrm{Soc}(M)) = \mathrm{Soc}(M)$.

The following notions may be considered an attempt to find complementary direct summands or to settle for a concept weaker than being a direct summand.

Definition 0.11. Let A be a submodule of M.

1) A submodule B of M is a **supplement** of A in M if $A + B = M$ and B is minimal with this property.
2) A submodule C of M is a **complement** of A in M if $A \cap C = 0$ and C is maximal with respect to this property. We also say that C is A–**high** in M.

A submodule is a complementary direct summand if and only if it is both a complement and a supplement. By Zorn's Lemma complements always exist but supplements need not exist. See [15, Section 5.2].

Definition 0.12.

1) A module P is **projective** if for every epimorphism $\varphi : M \to P$ there is a homomorphism $\psi : P \to M$ such that $\psi\varphi = 1_M$.
2) The module P is a **projective cover** of M if P is projective and there is an epimorphism $P \to M$ with small kernel.
3) A module Q is **injective** if for every monomorphism $\varphi : Q \to M$ there is a homomorphism $\psi : M \to Q$ such that $\psi\varphi = 1_Q$.
4) The module Q is an **injective hull** of M if Q is injective and there is a monomorphism $M \to Q$ with large image.

If $\varphi : M \to P$ is an epimorphism and P is projective, then $M \cong \mathrm{Ker}(\varphi) \oplus P$. Since every module is the epimorphic image of a free module, it follows that the projective modules are exactly the direct summands of free modules. If $\varphi : Q \to M$ is a monomorphism and Q is injective, then $M \cong Q \oplus \mathrm{Coker}(\varphi)$. Projective covers and injective hulls are unique up to isomorphism in a strong sense but projective covers may or may not exist while injective hulls always exist. See [15, Section 5.6].

Definition 0.13. Let M be an R–module.

1) The module M is **semiperfect** if every epimorphic image of M has a projective cover.
2) A ring R is (right) **perfect** if every right R–module has a projective cover.

On occasion we will need to use transfinite induction. It is generally accepted that the most useful variant in the context of algebra is the so–called **Zorn's Lemma**.

Lemma 0.14. *Let $\mathcal{S}$ be a family of subsets of a set S. If every ascending chain in $\mathcal{S}$ has an upper bound in $\mathcal{S}$, then $\mathcal{S}$ contains maximal elements.*

By $\mathbb{P}$ we denote the set of all prime numbers; $\mathbb{N}$, $\mathbb{Z}$, $\mathbb{Q}$ are the symbols for the sets of natural numbers (excluding 0), integers, and rational numbers respectively. Sometimes it is convenient to include 0 in the set of natural numbers and in this case we use the symbol $\mathbb{N}_0 := \mathbb{N} \cup \{0\}$.

Chapter II

Fundamental notions and properties

1 Partially invertible homomorphisms and the total

Let R be a ring with identity $1 \in R$, and denote by Mod–R the category of all right R–modules. In the following all modules are from Mod–R.

The next lemma gives information about f when f is a divisor of an idempotent.

Lemma 1.1. *For a mapping $f \in \mathrm{Hom}_R(M, W)$ the following statements are equivalent.*

1) *There exists $g \in \mathrm{Hom}_R(W, M)$ such that*

$$e := gf = e^2 \neq 0.$$

2) *There exists $h \in \mathrm{Hom}_R(W, M)$ such that*

$$d := fh = d^2 \neq 0.$$

3) *There exists $k \in \mathrm{Hom}_R(W, M)$ such that*

$$kfk = k \neq 0.$$

4) *There exist $0 \neq A \subseteq^{\oplus} M$, $B \subseteq^{\oplus} W$ such that the mapping*

$$A \ni a \mapsto f(a) \in B$$

is an isomorphism.

Proof. 1) $\Rightarrow$ 2) : Suppose that $e := gf = e^2 \neq 0$. Then also $e = egf$. Define

$$d := feg.$$

Then $(feg)(feg) = f(egfe)g = fe^3g = feg$, i.e., $d^2 = d$. Further, $gdf = gfegf = e^3 = e \neq 0$, hence $d \neq 0$. Take $h = eg$.

2) $\Rightarrow$ 1): Similar to the above. Define $e := hdf$. Then $e^2 = e \neq 0$.

1) $\Rightarrow$ 3): With e, g as in 1), it follows that

$$(eg)f(eg) = e^3g = eg,$$

and since $egf = e \neq 0$, also $eg \neq 0$. Set $k := eg$.

3) $\Rightarrow$ 1): From $kfk = k \neq 0$, it follows that $(kf)(kf) = kf$ and $kf \neq 0$. Set $e := kf$.

1) $\Rightarrow$ 4): By hypothesis $e = egf$ and $d = feg$, so $fe = fegf = df$. We will show that the mapping

$$\varphi : e(M) \ni e(m) \mapsto fe(m) \in d(W) \tag{1}$$

is an isomorphism. Since $fe(m) = df(m)$, φ is well–defined.

φ *is injective*: Assume $fe(m) = 0$. Then $0 = gfe(m) = e^2(m) = e(m)$.

φ *is surjective*: We have to show that, for $w \in W$, the element $d(w)$ is in the image of φ. For $g(w) \in M$ it follows that $fe(g(w)) = d(w)$.

4) $\Rightarrow$ 2): To prove this we need some notations. Let

$$M = A \oplus A_1, \qquad W = B \oplus B_1 \tag{2}$$

and denote by $\iota_A : A \to M$ and $\iota_B : B \to W$ the inclusions of A in M and of B in W, respectively. Further let $\pi_B : W \to B$ denote the projection of W onto B belonging to (2). We denote the isomorphism in 4) by φ, so $\varphi = \pi_B f \iota_A$. Let $h := \iota_A \varphi^{-1} \pi_B$. It then follows that

$$\begin{aligned}(fh)^2 &= f\iota_A\varphi^{-1}\pi_B f\iota_A\varphi^{-1}\pi_B \\ &= f\iota_A\varphi^{-1}(\pi_B f\iota_A)\varphi^{-1}\pi_B \\ &= f\iota_A\varphi^{-1}\varphi\varphi^{-1}\pi_B \\ &= f\iota_A\varphi^{-1}\pi_B = fh,\end{aligned}$$

hence $d := fh$ is an idempotent. From the definition of d it follows that

$$\pi_B d = \pi_B fh = \pi_B f\iota_A\varphi^{-1}\pi_B = (\pi_B f\iota_A)\varphi^{-1}\pi_B = \varphi\varphi^{-1}\pi_B = \pi_B.$$

With $A \neq 0$, also $B \neq 0$, and hence $\pi_B \neq 0$. This implies that $d \neq 0$. Therefore 2) is satisfied. □

Definition 1.2. We assume that the conditions of Lemma 1.1 are satisfied. Then

1) f is called **partially invertible** or pi for short;

2) g is called a **left partial inverse** of f and f is a **right partial inverse** of g;

3) the **total** of M to W is defined to be

$$\operatorname{Tot}(M, W) := \{f \in \operatorname{Hom}_R(M, W) \mid f \text{ is not pi.}\}.$$

The proof of Lemma 1.1 establishes the following remark.

Remark. Let $f \in \operatorname{Hom}_R(M, W)$ be pi, let $g \in \operatorname{Hom}_R(W, M)$ such that

$$0 \neq gf = e = e^2 \in \operatorname{End}(M).$$

Then the following hold.

1) For $d := feg \in \operatorname{End}_R(W)$, $0 \neq (fe)g = d = d^2 \in \operatorname{End}(W)$.
2) $M = e(M) \oplus (1 - e)(M), \quad W = d(W) \oplus (1 - d)(W)$.
3) $fe = df, \quad eg = gd$.
4) $\varphi : e(M) \ni x \mapsto f(x) \in d(W)$ is an isomorphism.
5) $\psi : d(W) \ni x \mapsto g(x) \in e(M)$ is an isomorphism.
6) $\operatorname{Ker}(f) \subseteq \operatorname{Ker}(e) = (1 - e)(M); \quad \operatorname{Ker}(g) \subseteq \operatorname{Ker}(d) = (1 - d)(W)$.
7) $\varphi\psi = 1_{d(W)}; \quad \psi\varphi = 1_{e(M)}$.
8) The map eg is both a left and a right partial inverse of f;
   the map df is both a left and a right partial inverse of g.

If $f \in \operatorname{Hom}_R(M, W)$ and if f is not pi, then by Lemma 1.1.4) f does not induce an isomorphism between any non–zero direct summands $A \subseteq^{\oplus} M$ and $B \subseteq^{\oplus} W$; therefore we also call such a map f a **total non–isomorphism**. The total is then the set of all total non–isomorphisms.

We call $f \in \operatorname{Hom}_R(W, M)$ **regular** if there exists $g \in \operatorname{Hom}_R(M, W)$ with $fgf = f$.

Lemma 1.3. *Let $f \in \operatorname{Hom}_R(M, W)$.*

1) *If $f \neq 0$ is regular, then f is partially invertible.*
2) *If f is partially invertible and $gf = e = e^2 \neq 0$, then fe and eg are non–zero regular homomorphisms.*

Proof. 1) Suppose that $f \neq 0$ is regular and $fgf = f$ for some $g \in \operatorname{Hom}_R(W, M)$. Then $f(gfg) = fg$ and $(fg)(fg) = (fgf)g = fg \neq 0$.

2) $e = egfe$ and so $(fe)g(fe) = fe$, $(eg)f(eg) = eg$. □

To establish the concepts of "partially invertible" and "total" also for a ring R, we identify R with $\operatorname{End}_R(R) = \operatorname{Hom}_R(R_R, R_R)$. Corresponding to every $r \in R$ there is the endomorphism

$$R_R \ni x \mapsto rx \in R_R.$$

For the sake of completeness we restate Lemma 1.1 for a ring R (without proof).

Lemma 1.4. *The following statements are equivalent for $r \in R$.*

1) *There exists $s \in R$ such that*

$$e := sr = e^2 \neq 0.$$

2) *There exists $t \in R$ such that*

$$d = rt = d^2 \neq 0.$$

3) *There exists $u \in R$ such that*

$$uru = u \neq 0.$$

4) *There exist $0 \neq A \subseteq^{\oplus} R_R$, $B \subseteq^{\oplus} R_R$ such that the mapping*

$$A \ni a \mapsto ra \in B$$

is an isomorphism.

The element $r \in R$ of the proper kind is also called **partially invertible** or pi, s is a **partial left inverse** of r, t is a partial right inverse of r, and the **total** of R is

$$\mathrm{Tot}(R) = \{r \in R \mid r \text{ is not pi}\}.$$

All that we will state for $\mathrm{Hom}_R(M, W)$ in the following can be specialized to R.

Recall that an element a of a ring R is regular if there is an element $x \in R$ such that $axa = a$. The ring R is **regular** if and only if every element of R is regular. By Lemma 1.3 if R is regular, then $\mathrm{Tot}(R) = 0$.

In [8, Theorem 1] it is shown that every ring contains a unique largest regular ideal $M(R)$. Furthermore, [8, Theorem 7] contains the following appealing result.

Let R be a ring satisfying the descending chain condition on right ideals. Then R is the ring direct product $R = M(R) \times M^*$ where M^* is the ideal consisting of all elements a of R such that $aM(R) = M(R)a = 0$.

It is easy to see that $\mathrm{Tot}(R) \subseteq M^*$. In fact, let $x \in \mathrm{Tot}(R)$ and $y \in M(R)$. Suppose that $yx \neq 0$. As $M(R)$ is an ideal, $yx \in M(R)$ and hence is regular and therefore pi. But then x is pi by Lemma 1.9, a contradiction. But $\mathrm{Tot}(R)$ cannot be equal to M^* since the latter contains a multiplicative identity if $1 \in R$. Brown–McCoy deal with rings that need not contain a multiplicative identity while we only consider rings with 1. We do not know any non–regular rings R with $M(R) \neq 0$.

Three easy examples will illustrate what the total can look like.

Example 1.5. Suppose that R is a ring that is directly indecomposable as a right R–module. Then $\mathrm{Tot}(R)$ is the complement of the group of units of R.

Proof. The only idempotents of R are 1 and 0 and hence the partially invertible elements of R are the invertible elements of R. □

Example 1.6. Suppose that M is a directly indecomposable module. Then $\mathrm{Tot}(\mathrm{End}(M)) = \mathrm{End}(M) \setminus \mathrm{Aut}(M)$.

Example 1.7. Let M and W be indecomposable R–modules that are not isomorphic. Then $\mathrm{Tot}(M, W) = \mathrm{Hom}_R(M, W)$.

Proof. The only non–zero direct summand of M is M itself, and the only non–zero direct summand of W is W. Consequently, a pi homomorphism must map M isomorphically onto W, but this is excluded by hypothesis. □

We return to the general situation. For future use we note a simple fact.

Proposition 1.8. *Let $R = R_1 \times \cdots \times R_n$ be a ring direct product. Then*

$$f = (f_1, \dots, f_n) \in \mathrm{Tot}(R) \Leftrightarrow \forall i, f_i \in \mathrm{Tot}(R_i).$$

Proof. Suppose that $f = (f_1, \dots, f_n) \in R$ is pi. Then there exist $e = (e_1, \dots, e_n) \in R$ and $g = (g_1, \dots, g_n) \in R$ such that $0 \neq fg = e = e^2$, i.e., for all i, $f_i g_i = e_i = e_i^2$ in R_i. Also $e \neq 0$ if and only if there exists i such that $e_i \neq 0$. Hence at least one of the f_i is pi. Conversely, if f_i is pi, say $0 \neq f_i g_i = e_i = e_i^2$ in R, then let $e = (0, \dots, e_i, \dots, 0)$ and $g = (0, \dots, g_i, \dots, 0)$. Then $0 \neq fg = e = e^2$, showing that f is pi in R. We have shown that f is pi in R if and only if there exists i such that f_i is pi in R_i. The negation of this statement is the claim of the proposition. □

As can be seen from the examples, the total of a ring is, in general, not an ideal (closed under addition and right and left multiplication), but it is a **semi–ideal** in Mod–R, i.e., closed under right and left multiplication. This is an easy consequence of the following Lemma 1.9.

Lemma 1.9. *If a product of homomorphisms in* Mod–R *is partially invertible, then each of its factors is partially invertible.*

Proof. Induction on the number n of factors of the product.

$n = 2$. Assume that $f_1 f_2$ is pi. Then there are equations

$$e = g(f_1 f_2) = (g f_1) f_2, \qquad d = (f_1 f_2) h = f_1 (f_2 h).$$

By Lemma 1.1.1) f_2 is pi and by Lemma 1.1.2) f_1 is pi.

Induction step. Assume that the assertion is true for $n \geq 2$ and $f_1 f_2 \cdots f_n f_{n+1}$ is pi. Then by the case $n = 2$, $f_1 f_2 \cdots f_n$ and f_{n+1} are pi and then, by induction hypothesis, each of $f_1, f_2, \dots, f_n$ is pi. □

Corollary 1.10. *For arbitrary $M, W, X, Y \in$*Mod–R *we have*

$$\begin{aligned} &\forall f \in \mathrm{Hom}_R(W, Y), \forall h \in \mathrm{Hom}_R(X, M), \\ &f\,\mathrm{Tot}(M, W) h \subseteq \mathrm{Tot}(X, Y). \end{aligned} \tag{3}$$

Proof. A product fgh in (3) cannot be pi since otherwise g would be pi. □

For arbitrary $M, W \in$ Mod–R, $0 \in \mathrm{Tot}(M, W)$, hence the total is not empty. We use (3) as the defining property for **semi–ideal** in Mod–R, i.e., a semi–ideal is a subset of some group $\mathrm{Hom}_R(M, W)$ satisfying (3).

Some light can be shed onto the structure of the total with the following observation. We use that $\mathrm{Hom}_R(M, W)$ is an $\mathrm{End}(W)$–$\mathrm{End}(M)$ bimodule and by a submodule of $\mathrm{Tot}(M, W)$ we mean a submodule of $\mathrm{Hom}_R(M, W)$ that happens to be contained in $\mathrm{Tot}(M, W)$.

Proposition 1.11. *Let $M, W \in$Mod–R. Then the total* $\mathrm{Tot}(M, W)$ *is the set–theoretic union of cyclic left* $\mathrm{End}(W)$*–submodules and also the set–theoretic union of cyclic right* $\mathrm{End}(M)$*–submodules.*

Proof. Let $g \in \mathrm{Tot}(M, W)$. Then $\mathrm{End}(W)g = \{fg \mid f \in \mathrm{End}(W)\}$ is a cyclic left $\mathrm{End}(W)$–module that is contained in $\mathrm{Tot}(M, W)$ due to Corollary 1.10. The total is the union of all submodules $\mathrm{End}(W)g$. The other case is analogous. □

In an appendix to this chapter we will discuss further the structure of semi–ideals.

Since, in general, the total is not an ideal in Mod–R, i.e., $\mathrm{Tot}(M, W)$ is not additively closed for all $M, W \in$ Mod–R, there is the natural question: Do there exist rings R such that the total is a non–zero ideal in Mod–R, i.e., do there exist rings R such that $\mathrm{Tot}(M, W)$ is additively closed for any two modules M, W in Mod–R and in some cases $\mathrm{Tot}(M, W) \neq 0$? For any semisimple ring R the total is the zero–ideal of Mod–R (Example 1.12).

Example 1.12. Let M and W be semisimple R–modules. Then $\mathrm{Tot}(M, W) = 0$. If R is semi–simple, then the total is the zero–ideal in Mod–R.

Proof. Since M, W are semisimple, every submodule of M is a direct summand and every submodule of W is a direct summand. Let $0 \neq f \in \mathrm{Hom}_R(M, W)$. Then $M = A \oplus \mathrm{Ker}(f)$ and $W = \mathrm{Im}(f) \oplus B$ for some modules $A \neq 0$, B; moreover, f induces an isomorphism

$$A \ni a \mapsto f(a) \in \mathrm{Im}(f).$$

Therefore, (Lemma 1.1.4)) every $0 \neq f \in \mathrm{Hom}_R(M, W)$ is pi and $\mathrm{Tot}(M, W) = 0$.

Assume that R is a semisimple ring. Then all R–modules are semisimple and the total is the zero–ideal in the category. □

The other extreme is illustrated by the ring $\mathbb{Z}$ which is a concrete case of Example 1.5.

Example 1.13. $\mathrm{Tot}(\mathbb{Z}) = \mathbb{Z} \setminus \{1, -1\} = \bigcup_{p \in \mathbb{P}} p\mathbb{Z}$, and $\mathrm{Tot}(\mathbb{Z})$ is not closed under addition, hence not an ideal.

Later (see III.Corollary 3.2) we will see that $\mathrm{Rad}(R) = \mathrm{Tot}(R)$ holds for Artinian rings R. Since $\mathrm{Rad}(R)$ is additively closed, the same is true for $\mathrm{Tot}(R)$.

We also have a certain general additive closure property: The sum of an element in the radical with one in the total is again in the total (Theorem 2.4).

2 The connection between the radical and the total

Set

$$S := \mathrm{End}(W), \quad T := \mathrm{End}(M).$$

Then $\mathrm{Hom}_R(M,W)$ is an S–T–bimodule. We have the radical of the left S–module ${}_S\mathrm{Hom}_R(M,W)$ and denote it by $\mathrm{Rad}({}_S\mathrm{Hom}_R(M,W))$, and that of the right T–module $\mathrm{Hom}_R(M,W)_T$ which we denote by $\mathrm{Rad}(\mathrm{Hom}_R(M,W)_T)$. We will introduce a third radical of $\mathrm{Hom}_R(M,W)$ denoted $\mathrm{Rad}(M,W)$. This requires some well–known facts about radicals. For completeness we include them with proofs.

Lemma 2.1. *Let R be a ring with 1.*

1) *Let A be a non–void subset of R that is closed under right multiplication by elements of R. Then $A \subseteq \mathrm{Rad}(R)$ if and only if for all $a \in A$, $1-a$ is an invertible element in R.*

2) *Let $f \in \mathrm{Hom}_R(M,W)$ and $g \in \mathrm{Hom}_R(W,M)$. Then $1_S - fg$ is invertible in $S = \mathrm{End}_R(W)$ if and only if $1_T - gf$ is invertible in $T = \mathrm{End}_R(M)$.*

3) $\{f \in \mathrm{Hom}_R(M,W) \mid \forall\, g \in \mathrm{Hom}_R(W,M), gf \in \mathrm{Rad}(T)\}$
$= \{f \in \mathrm{Hom}_R(M,W) \mid \forall\, g \in \mathrm{Hom}_R(W,M), fg \in \mathrm{Rad}(S)\}$.

Proof. 1) Assume first that $A \subseteq \mathrm{Rad}(R)$. Then, for every $a \in A$, $aR \subseteq^\circ R_R$, and hence $(1-a)R + aR = R$ implies that $(1-a)R = R$. So there is $s \in R$ such that $(1-a)s = 1$ and so $1 + as = s$. Since also $a(-s) \in A$, by specialization, there exists $t \in R$ with $(1+as)t = st = 1$. It follows that $1 - a = (1-a)st = ((1-a)s)t = t$ and further $1 = st = s(1-a)$. Hence s is the inverse of $1-a$.

Conversely, assume that for every $a \in A$, $1-a$ is an invertible element in R. Denote by $|A)$ the right ideal generated by A. We will show that $|A) \subseteq^\circ R_R$ which means that $|A) \subseteq \mathrm{Rad}(R)$. Assume that B is a right ideal of R with $R = B + |A)$. Then

$$1 = b + a_1 + \cdots + a_m, \quad b \in B,\ a_i \in A. \tag{4}$$

If $1 = b \in B$, then $B = R$ and we are done. Therefore we can assume that a_m with $m \geq 1$ really occurs. By assumption,

$$1 - a_m = b + a_1 + \cdots + a_{m-1}$$

is invertible, hence

$$1 = (b + a_1 + \cdots + a_{m-1})(1 - a_m)^{-1}. \tag{5}$$

Since $b(1-a_m)^{-1} \in B$ and $a_i(1-a_m)^{-1} \in A$ for $i = 1, \ldots, m-1$, we have in (5) a representation analogous to (4) but with only $m-1$ or no (for $m=1$) summands from A. By induction we get that $1 \in B$ and $|A) \subseteq \mathrm{Rad}(R)$.

2) Suppose that $1_S - fg$ is invertible in S. It is easy to check that

$$(1_T - gf)^{-1} = 1_T + g(1_S - fg)^{-1} f.$$

Similarly, if $1_T - gf$ is invertible in T, then

$$(1_S - fg)^{-1} = 1_S + f(1_T - gf)^{-1}g.$$

3) Suppose that $f \in \mathrm{Hom}_R(M,W)$ and for every $g \in \mathrm{Hom}_R(W,M)$, $fg \in \mathrm{Rad}(S)$. Since for all $s \in S$, $t \in T$, $g \in \mathrm{Hom}_R(W,M)$, also $tgs \in \mathrm{Hom}_R(W,M)$, and so $ftgs \in \mathrm{Rad}(S)$. Hence the right ideal $ftgS$ is contained in $\mathrm{Rad}(S)$. By Lemma 2.1.1) it follows that $1_S - ftgs$ is invertible in S. Then, by Lemma 2.1.2), $Tgsf$ is a left ideal contained in $\mathrm{Rad}(T)$. Therefore $f \in \{f \in \mathrm{Hom}_R(M,W) \mid \forall\, g \in \mathrm{Hom}_R(W,M), gf \in \mathrm{Rad}(T)\}$. The reverse containment follows in a similar fashion. □

On the basis of Lemma 2.1, we can define the announced radical.

Definition 2.2.

$$\begin{aligned}\mathrm{Rad}(M,W) &:= \{f \in \mathrm{Hom}_R(M,W) \mid \forall\, g \in \mathrm{Hom}_R(W,M), fg \in \mathrm{Rad}(S)\}\\ &= \{f \in \mathrm{Hom}_R(M,W) \mid \forall\, g \in \mathrm{Hom}_R(W,M), gf \in \mathrm{Rad}(T)\}.\end{aligned}$$

For a ring R identified naturally with $\mathrm{Hom}_R(R_R, R_R)$ we obtain the usual radical: $\mathrm{Rad}(R_R, R_R) = \mathrm{Rad}(R)$.

We return to Lemma 2.1.1) with a remark. We did not assume that A is a right ideal and later will apply Lemma 2.1.1) to a right semi–ideal.

Corollary 2.3. $\mathrm{Rad}({}_S\mathrm{Hom}_R(M,W)) \cup \mathrm{Rad}(\mathrm{Hom}_R(M,W)_T) \subseteq \mathrm{Rad}(M,W)$.

Proof. Let $f \in \mathrm{Rad}({}_S\mathrm{Hom}_R(M,W))$, which means that

$$Sf \subseteq^\circ {}_S\mathrm{Hom}_R(M,W).$$

Then, for any $g \in \mathrm{Hom}_R(W,M)$, right multiplication by g is an S–homomorphism and small submodules are mapped to small submodules under homomorphisms, so also

$$Sfg \subseteq^\circ {}_SS,$$

hence $fg \in \mathrm{Rad}(S)$. This implies that $f \in \mathrm{Rad}(M,W)$. The other case is similar. □

In spite of the fact that the total is in general not an ideal, it always has the following interesting additive closure property.

Theorem 2.4. *For arbitrary $M, W \in$ Mod–R,*

$$\mathrm{Rad}(M,W) + \mathrm{Tot}(M,W) = \mathrm{Tot}(M,W). \tag{6}$$

Proof. Indirect. Let $f \in \mathrm{Rad}(M,W)$, $g \in \mathrm{Tot}(M,W)$ and assume that $f+g$ is pi. Then there exists $h \in \mathrm{Hom}_R(W,M)$ such that

$$e := (f+g)h = e^2 \neq 0, e \in S.$$

Hence $e = fh + gh$ with $fh \in \mathrm{Rad}(S)$. It follows, using $fhS \subseteq^\circ S_S$, that

$$S = eS \oplus (1-e)S = fhS + ghS + (1-e)S = ghS + (1-e)S.$$

This implies that $eS = eghS$. Hence there exists $s \in S$ with $e = eghs$. Since e is pi, the element $eghs$ is also pi, and by Lemma 1.9, g is pi, a contradiction. □

Since $0 \in \mathrm{Tot}(M, W)$, (6) implies that $\mathrm{Rad}(M, W) \subseteq \mathrm{Tot}(M, W)$. But even more is contained in the total.
If we write

$$\Delta(M, W) := \{f \in \mathrm{Hom}_R(M, W) \mid \mathrm{Ker}(f) \subseteq^* M\}$$

and

$$\nabla(M, W) := \{f \in \mathrm{Hom}_R(M, W) \mid \mathrm{Im}(f) \subseteq^\circ W\}$$

for the **singular** respectively **co–singular** submodule of $\mathrm{Hom}_R(M, W)$, then also

$$\Delta(M, W) \cup \nabla(M, W) \subseteq \mathrm{Tot}(M, W). \tag{7}$$

To see these, assume to the contrary that $f \in \Delta(M, W)$ is pi. Then, if

$$e := gf = e^2 \neq 0,$$

then e must also have a large kernel. Since the kernel of e is $(1-e)T$, and since this is a direct summand of T, it must equal T which implies that $e = 0$, a contradiction. Similarly, if $f \in \nabla(M, W)$ and

$$d := fh = d^2 \neq 0,$$

then $\mathrm{Im}(d) \subseteq \mathrm{Im}(f) \subseteq^\circ W$. But the image of the idempotent $d \neq 0$ that is the direct summand dS, is not small in S, again a contradiction. □

Since Rad, Δ, ∇ are all contained in the total, the question arises: for which modules is any of these equal to the total? Rad, Δ, ∇ are all additively closed; hence, if one of these equals the total, then this total is an ideal in Mod–R.

The question of equality will be the main concern of the next chapter. Here we will deal with the special case Rad = Tot for which we later have an interesting application (Corollary 2.8) that in turn will be used in Chapter V in the context of LE–decompositions.

Theorem 2.5. *Suppose that* $M = A \oplus B \in$ Mod–R. *Set* $S := \mathrm{End}_R(M)$ *and* $T := \mathrm{End}_R(A)$. *Further let* $0 \neq e = e^2 \in S$ *be the projector onto* A *along* B.

1) *If* $\mathrm{Rad}(T) = \mathrm{Tot}(T)$, *then*

$$e\,\mathrm{Tot}(S) \subseteq \mathrm{Rad}(S) \quad \text{and} \quad \mathrm{Tot}(S)e \subseteq \mathrm{Rad}(S).$$

2) *If* $\mathrm{Tot}(S)e \subseteq \mathrm{Rad}(S)$, *then, for all* $f \in \mathrm{Tot}(S)$ *and all* $0 \neq a \in A$, *it is true that* $(1_M - f)(a) \neq 0$.

Proof. Let

- $\iota :=$ inclusion of A in M,
- $\pi :=$ projection of M onto A along B.

Then

$$e = \iota\pi, \quad 1_A = \iota\pi, \quad \pi = \pi e, \quad \iota = e\iota,$$

which will be used in the following.

1) We apply Lemma 2.1.1) to the semi–right–ideal $e\,\mathrm{Tot}(S)$ and the semi–left–ideal $\mathrm{Tot}(S)e$. Hence it suffices to show that for $f \in \mathrm{Tot}(S)$, the maps $1_M - ef$ and $1_M - fe$ are invertible in S. By Lemma 2.1.2), $1_M - ef = 1_M - e(ef)$ is invertible if and only if $1_M - efe$ is invertible. Similarly, $1_M - fe$ is invertible if and only if $1_M - efe$. It is clear that eSe is a ring with identity element e. Since $\mathrm{Tot}(S)$ is a semi–ideal, it follows that $\pi f\iota \in \mathrm{Tot}(T) = \mathrm{Rad}(T)$. Consider the following homomorphisms

$$T \ni t \mapsto \iota t\pi = e(\iota t\pi)e \in eSe, \tag{8}$$

$$eSe \ni ege \mapsto \pi(ege)\iota \in T. \tag{9}$$

It is easy to see that these are inverses of each other, hence ring isomorphisms. Then by (8), the radical of T is mapped onto the radical of eSe. This means, for $f \in \mathrm{Tot}(S)$, $\pi f\iota \in T$, that $efe \in \mathrm{Rad}(eSe)$. So $e - efe$ is invertible in eSe. Denote its inverse by ege. Then

$$(e - efe)ege = e = ege(e - efe).$$

With this, we obtain for $1_M - efe$ that

$$\begin{aligned}(1_M - efe)(1_M - e + ege) &= ((1_M - e) + (e - efe))((1_M - e) + ege)\\ &= 1_M - e + e = 1_M,\end{aligned}$$

and also

$$\begin{aligned}(1_M - e + ege)(1_M - efe) &= ((1_M - e) + ege)((1_M - e) + e - efe)\\ &= 1_M - e + e = 1_M.\end{aligned}$$

As mentioned at the beginning of this proof, then also $1_M - ef$ and $1_M - fe$ have inverses in S, which was to be shown.

2) Now let $f \in \mathrm{Tot}(S)$ and $0 \neq a \in A$. Then

$$(1_M - f)(a) = a - f(a) = a - f(ea) = a - (fe)(a) = (1_M - fe)(a).$$

By assumption $fe \in \mathrm{Rad}(S)$, hence $1_M - fe$ is invertible. Therefore

$$\begin{aligned}a &= (1_M - fe)^{-1}(1_M - fe)(a) = (1_M - fe)^{-1}(a - fe(a))\\ &= (1_M - fe)^{-1}(1_M - f)(a).\end{aligned}$$

If $(1_M - f)(a) = 0$, then $a = 0$. Hence, a being non–zero, $(1_M - f)(a) \neq 0$. $\square$

Definition 2.6. A decomposition

$$M = \bigoplus_{i\in I} M_i \tag{10}$$

is called an **RT–decomposition** if and only if, for all $i \in I$,

$$\text{Rad}(\text{End}(M_i)) = \text{Tot}(\text{End}(M_i)).$$

(R stands for radical and T stands for total.)

Later, we will see that decompositions (10) with the following properties are RT–decompositions.

1) The endomorphism rings of all M_i are local. In this case (10) is called an **LE–decomposition**.

2) All modules M_i are injective.

Question 2.7. Under which conditions is it true that $\text{Rad}(\text{End}(M)) = \text{Tot}(\text{End}(M))$ for a decomposition (10)? In Chapter IV we answer this question for LE–decompositions (IV.4.1).

Corollary 2.8. *Assume that* (10) *is an RT–decomposition. Then the following statements hold.*

1) *For every* e_i, $i \in I$,

$$e_i \,\text{Tot}(S) \subseteq \text{Rad}(S), \quad \text{Tot}(S)e_i \subseteq \text{Rad}(S).$$

2) *For every* $f \in \text{Tot}(S)$, $1_M - f$ *is injective.*

Proof. 1) Let $A = M_i$ in Theorem 2.5.1) and the claim follows.

2) Let $0 \neq a \in M$ and set $d := \sum_{i\in \text{spt}(a)} e_i$. Assume that $f \in \text{Tot}(S)$. Then

$$(1_M - f)(a) = a - f(a) = a - fd(a) = (1_m - fd)(a).$$

Since $fe_i \in \text{Rad}(S)$ and $\text{Rad}(S)$ is an ideal, also $fd \in \text{Rad}(S)$. Since $0 \neq a \in d(M)$, we can apply Theorem 2.5.2) with $A := d(M)$ to obtain the claim. □

3 Connection with regularity, decomposition theorems

If R is a regular ring, then every non–zero element of R is partially invertible, hence $\text{Tot}(R) = 0$. It is natural to ask whether $\text{Tot}(R) = 0$ implies that the ring R is regular. The answer is no! (See Example 3.5) However there are interesting connections between partial invertibility and regularity not only for rings but for homomorphisms of modules. We will next derive results that connect regularity and partial invertibility.

In what follows we have to repeat partly the proof of Lemma 1.1 but with added details. For the convenience of the reader we will give the complete proof without reference to Lemma 1.1.

We again use the notation

$$S := \operatorname{End}(W), \quad T := \operatorname{End}(M),$$

so that $\operatorname{Hom}_R(M, W)$ is an S–T–bimodule.

Lemma 3.1. *Assume that $f \in \operatorname{Hom}_R(M, W)$ is partially invertible and for some $g \in \operatorname{Hom}_R(W, M)$ we have $0 \neq gf =: e = e^2 \in T$. Then the following statements hold.*

1) *The mappings*

$$eT \ni et \mapsto fet \in feT, \quad Te \ni te \mapsto teg \in Teg,$$

are isomorphisms, hence feT and Teg are projective submodules of $\operatorname{Hom}_R(M, W)_T$ and ${}_T\operatorname{Hom}_R(W, M)$ respectively.

2)

$$\operatorname{Hom}_R(M, W)_T = feT \oplus \operatorname{Ker}(feg-),$$
$${}_T\operatorname{Hom}_R(W, M) = Teg \oplus \operatorname{Ker}(-feg),$$

where $feg-$ and $-feg$ are the homomorphisms

$$feg- \ : \ \operatorname{Hom}_R(M, W)_T \ni h \mapsto fegh \in feT,$$
$$-feg \ : \ {}_T\operatorname{Hom}_R(W, M) \ni h \mapsto hfeg \in Teg.$$

Proof. 1) It is only necessary to show that the mappings are injective. Assume that $fet = 0$. Then it follows that $et = e^2t = gfet = g0 = 0$. If $teg = 0$, then $te = te^2 = tegf = 0$. Since, respectively, eT and Te are direct summands of T_T and ${}_TT$, they are projective and so are feT and Teg.

2) We consider the commutative diagram

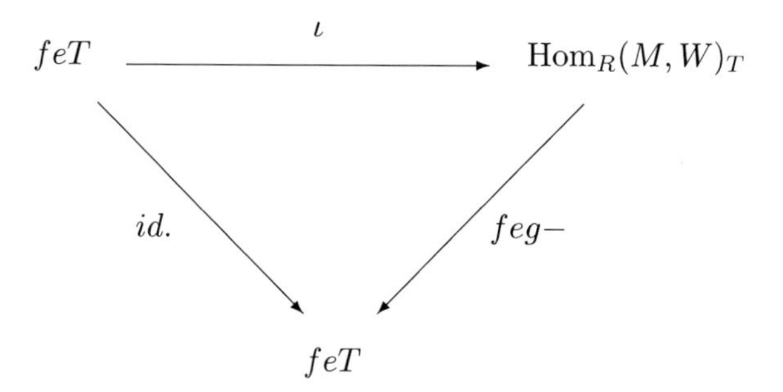

where ι is the inclusion. Then by 1), $feg(fet) = fet$ for all $t \in T$, hence feg is the identity map id. It follows by Lemma 0.2 (with $gf = 1$) that

$$\mathrm{Hom}_R(M, W)_T = \mathrm{Im}(\iota) \oplus \mathrm{Ker}(feg-) = feT \oplus \mathrm{Ker}(feg-).$$

The proof for ${}_T\,\mathrm{Hom}_R(W, M)$ is similar. □

In later applications the following fact that was established in the proof will have to be used repeatedly.

Remark. If $f \in \mathrm{Hom}_R(M, W)_T$ is partially invertible, then the regular homomorphism fe belongs to fT. Hence $feT \subseteq fT$.

The following decomposition theorem shows that the notions "partially invertible" and "total" are useful and interesting in connection with regularity.

Theorem 3.2. *For arbitrary modules M, W, and $T = \mathrm{End}_R(W)$, one of the following cases occurs.*

1) $\mathrm{Hom}_R(M, W) = \mathrm{Tot}(M, W)$.
2) $\mathrm{Hom}_R(M, W) = \bigoplus_{i=1}^n f_i T \oplus U$ *with* $n \geq 1$, $U \subseteq \mathrm{Tot}(M, W)$, $f_i \in \mathrm{Hom}_R(M, W)$ *regular, and $f_i T$ projective (and having the further properties of* Lemma 3.1*) for $i = 1, \dots, n$.*
3) $\mathrm{Hom}_R(M, W)$ *contains an independent family of the form $\{f_i T \mid i \in \mathbb{N}\}$ that finitely generates direct summands where the $f_i T$ share the properties of the $f_i T$ in* 2).

Proof. We assume that 1) is not satisfied. Then there exists $f \in \mathrm{Hom}_R(M, W)$ that is pi. Let f be such a mapping. We apply Lemma 3.1 to this f and set $f_1 := fe$. Then Lemma 3.1.3) implies that

$$\mathrm{Hom}_R(M, W) = f_1 T \oplus B_1, \text{ where } B_1 := \mathrm{Ker}(f_1 g_1 -). \tag{11}$$

If $B_1 \subseteq \mathrm{Tot}(M, W)$, then we have 2) with $n = 1$ and $U := B_1$. If $B_1 \nsubseteq \mathrm{Tot}(M, W)$, then there exists a pi homomorphism in B_1 to which we apply the foregoing construction to get $\mathrm{Hom}_R(M, W) = f_2 T \oplus B_2$ with $f_2 T \subseteq B_1$, so

$$B_1 = f_2 T \oplus (B_1 \cap B_2),$$

and using (11)

$$\mathrm{Hom}_R(M, W) = f_1 T \oplus f_2 T \oplus (B_1 \cap B_2).$$

If $B_1 \cap B_2 \subseteq \mathrm{Tot}(M, W)$, then we are finished. Otherwise we continue inductively. Either the induction stops and we have case 2), or it continues without end and results in case 3) In 3) "finitely generates direct summands" means that for any finite set $I \subseteq \mathbb{N}$, the sub-sum $\bigoplus_{i \in I} f_i T$ is a direct summand of $\mathrm{Hom}_R(M, W)_T$. This is clear since by the induction process

$$\mathrm{Hom}_R(M, W) = \bigoplus_{i=1}^n f_i T \oplus \left(\bigcap_{i=1}^n B_i\right).$$

(Take $n := \max\{i \mid i \in I\}$.) □

If $\mathrm{Tot}(M, W) \neq \mathrm{Hom}_R(M, W)$, then we have seen that there exist at least one $f_i T$ with the properties in Lemma 3.1. We consider now an independent family $\{f_i T \mid i \in I\}$ where the f_i again have the properties of Lemma 3.1. "Independent" means that

$$\sum_{i \in I} f_i T = \bigoplus_{i \in I} f_i T.$$

Obviously, to such families we can apply Zorn's Lemma (the union of an ascending chain of such families is again such a family). Therefore there exists a maximal independent such family $\{f_j T \mid j \in J\}$.

Remark. If $\{f_j T \mid j \in J\}$ is a maximal independent family where the $f_j T$ have the properties of Lemma 3.1, then the following statements hold.

1) If $A \subseteq \mathrm{Hom}_R(M, W)_T$ and if $\left(\bigoplus_{j \in J} f_j T\right) \cap A = 0$, then $A \subseteq \mathrm{Tot}(M, W)$.

2) If $\mathrm{Tot}(M, W) = 0$, then $\bigoplus_{j \in J} f_j T$ is large in $\mathrm{Hom}_R(M, W)_T$.

Proof. 1) If we assume that $A \not\subseteq \mathrm{Tot}(M, W)$, then A contains a pi homomorphism f and thus the regular homomorphism $fe \in A$. But this would mean that the family $\{f_j \mid j \in J\}$ is not maximal independent, a contradiction.

2) follows from 1) □

Since the sum $\bigoplus_{j \in J} f_j T$ of the projective modules $f_j T$ is again projective, the Remark means that $\mathrm{Hom}_R(M, W)_T$ is "nearly" projective (in the sense of the above Remark).

For the next decomposition theorem, we need some more technical details that we will collect.

Assume that $f \in \mathrm{Hom}_R(M, W)$, $g \in \mathrm{Hom}_R(W, M)$ with

$$e := gf = e^2 \neq 0.$$

We replace g by eg and denote it again by g. Then $eg = g$.

Then define $d := fg$ which implies

$$d^2 = fgfg = feg = fg = d, \quad gdf = gfgf = e^2 = e \neq 0, \quad d \neq 0.$$

Now consider the mappings

$$\begin{aligned} e(M) \ni e(x) &\mapsto fe(x) \in d(W), \\ d(W) \ni d(y) &\mapsto gd(y) \in e(M), \end{aligned} \tag{12}$$

which we will show to be isomorphisms and inverse to each other.

We will show further that

$$f(1-e)(M) \subseteq (1-d)(W), \tag{13}$$

$$g(1-d)(W) \subseteq (1-e)(M), \tag{14}$$

$$\mathrm{Ker}(f) \subseteq (1-e)(M), \qquad \mathrm{Ker}(g) \subseteq (1-d)(W). \tag{15}$$

For the proof of (13), (14), (15) we use the following equalities.

$$fe = fgf = df, \quad gd = gfg = eg(= g). \tag{16}$$

These imply

$$fe(M) = df(M) \subseteq d(W), \quad gd(W) = eg(W) \subseteq e(M),$$

and we see that (12) are well–defined homomorphisms. Further, the compositions of the homomorphisms in (12)

$$e(x) \mapsto fe(x) \mapsto gfe(x) = e^2(x) = e(x)$$

and

$$d(y) \mapsto gd(y) \mapsto fgd(y) = d^2(y) = d(y)$$

are the identities. Hence these maps are isomorphisms, inverse to each other. Now from (16) it follows that

$$f(1-e) = (1-d)f, \quad g(1-d) = (1-e)g$$

and these equalities imply (13) and (14). To prove (15) let $x \in \mathrm{Ker}(f)$. Then from $f(x) = 0$ it follows with (16) that

$$0 = df(x) = fe(x).$$

The first isomorphism in (12) then implies that $e(x) = 0$, hence

$$x = e(x) + (1-e)(x) = (1-e)(x) \subseteq (1-e)(M),$$

that is, $\mathrm{Ker}(f) \subseteq (1-e)(M)$. Similarly, $\mathrm{Ker}(g) \subseteq (1-d)(W)$. Hence (15) is proved.

The best scenario for $\mathrm{Hom}_R(M, W)$ occurs if $\mathrm{Tot}(M, W) = 0$. In this case we obtain interesting information about the homomorphisms $f : M \to W$.

Theorem 3.3. *Suppose that* $\mathrm{Tot}(M, W) = 0$ *and* $0 \neq f \in \mathrm{Hom}_R(M, W)$. *Then one of the following cases occurs.*

Case 1. *There exist decompositions*

$$M = M_1 \oplus \mathrm{Ker}(f), \quad W = \mathrm{Im}(f) \oplus D_1$$

and f *induces an isomorphism*

$$f_1 : M_1 \ni x \mapsto f(x) \in \mathrm{Im}(f).$$

Furthermore, f *is regular.*

Case 2. *There exists a sequence of decompositions*

$$M = M_n \oplus B_n, \quad W = W_n \oplus D_n, \quad n \in \mathbb{N},$$

with the following properties.

i) *f induces isomorphisms* $f_n : M_n \ni x \mapsto f(x) \in W_n$,

ii) $f(B_n) \subseteq D_n, \;\; \mathrm{Ker}(f) \subseteq B_n$,

iii) $M_n \subsetneqq M_{n+1}, \; W_n \subsetneqq W_{n+1}, \; B_{n+1} \subsetneqq B_n, \; D_{n+1} \subsetneqq D_n$.

Proof. We construct inductively the modules M_n, B_n, W_n and D_n. The first case occurs if the constructions stabilize.

Start of Induction. Since f is pi, we have e and d with the properties (12) through (15). Thus, defining

$$M_1 := e(M), \;\; B_1 := (1-e)(M), \;\; W_1 := d(W), \;\; D_1 := (1-d)(W),$$

all claims are satisfied for $n = 1$ by (12) through (15). If $\mathrm{Ker}(f) = B_1$, then $\mathrm{Im}(f) = W_1$, and we have Case 1.

Induction from n to $n+1$. We assume now that Case 2 is true for n. Then the mapping

$$h : B_n \ni x \mapsto f(x) \in D_n$$

is non–zero because of the assumption that $\mathrm{Ker}(f) \subsetneqq B_n$. Denote by π the projection of $M = M_n \oplus B_n$ onto B_n along M_n, and by $\iota : D_n \to W$ the inclusion. Then $0 \neq \iota h \pi \in \mathrm{Hom}_R(M, W)$ and since $\mathrm{Tot}(M, W) = 0$, it follows that $\iota h \pi$ is pi and then also h is pi. We can now apply the beginning of the induction to h (in place of f), and obtain decompositions

$$B_n = A \oplus B_{n+1}, \quad D_n = C \oplus D_{n+1}$$

with $A \neq 0$ and the isomorphism

$$A \ni x \mapsto h(x) = f(x) \in C.$$

Furthermore, it follows that

$$h(B_{n+1}) = f(B_{n+1}) \subseteq D_{n+1}$$

and

$$\mathrm{Ker}(h) = \mathrm{Ker}(f) \subseteq B_{n+1}.$$

We define

$$M_{n+1} := M_n \oplus A, \quad W_{n+1} := W_n \oplus C$$

and

$$f_{n+1} : M_{n+1} \ni x \mapsto f(x) \in W_{n+1}.$$

If there exists an N with $\mathrm{Ker}(f) = B_N$ (in ii)), then $\mathrm{Im}(f) = W_N$ and we have the first case (but with $M_1 = M_N$, $\mathrm{Ker}(f) = B_N$, $\mathrm{Im}(f) = W_N$, $D_1 = D_N$). Otherwise the induction continues and Case 2 occurs.

We still have to prove that in Case 1, f is regular. Denote by π the projection of $W = \mathrm{Im}(f) \oplus D_1$ onto $\mathrm{Im}(f)$ along D_1, and by $\iota : M_1 \to M$ the inclusion. Then define $h := \iota f_1^{-1}\pi$. For $x \in M_1$ and $y \in \mathrm{Ker}(f)$ it follows that

$$fhf(x+y) = f\iota f_1^{-1}\pi f(x) = f\iota f_1^{-1}f(x) = f(x) = f(x+y),$$

hence $fhf = f$. The proof of Theorem 3.3 is now complete. □

Theorem 3.3 is also interesting in the special case of a ring R (which means $M = W = R$).

Corollary 3.4. *If* $\mathrm{Tot}(R) = 0$ *and if* $0 \neq f \in R$, *then one of the following cases occurs, where* $f \in R$ *is identified with left multiplication by* f *in* R_R.

Case 1. *There exist decompositions*

$$R_R = R_1 \oplus \mathrm{Ker}(f) = \mathrm{Im}(f) \oplus S_1$$

and f *induces an isomorphism*

$$f_1 : R_1 \ni x \mapsto f(x) \in \mathrm{Im}(f).$$

Furthermore, f *is regular.*

Case 2. *There exist sequences of decompositions*

$$R_R = R_n \oplus S_n = T_n \oplus U_n, \quad n \in \mathbb{N},$$

with the following properties.

i) f *induces isomorphisms* $f_n : R_n \ni x \mapsto f(x) \in T_n$,

ii) $f(S_n) \subseteq U_n$, $\mathrm{Ker}(f) \subseteq S_n$,

iii) $R_n \subsetneqq R_{n+1}$, $S_n \subsetneqq S_{n+1}$, $T_{n+1} \subsetneqq T_n$, $U_{n+1} \subsetneqq U_n$.

We are now ready to give an example of a non–regular ring R with $\mathrm{Tot}(R) = 0$.

Example 3.5. There exist rings with trivial total that are not regular.

Proof. Let $S := \mathbb{Q}^{\mathbb{N}}$ with component–wise addition and multiplication. Denote the elements of S by (q_i). We consider the subring R of S consisting of all elements (q_i) for which there is an integer n (depending on (q_i)) with $q_i \in \mathbb{Z}$ for all $i \geq n$. The idempotents in R are the (q_i) with $q_i \in \{0, 1\}$ for all $i \in \mathbb{N}$. Then obviously $\mathrm{Tot}(R) = 0$. But R is not regular; for example, the element $(2, 2, 2, \ldots)$ is not regular in R. □

One more remark on the general situation. If, in Theorem 3.3, M or W satisfies the chain condition for direct summands, then Case 2 cannot occur and $\mathrm{Hom}_R(M, W)$ is regular.

4 The dual case; the total of a module

We consider here a special case that is interesting in itself but which is also connected with other notions and results.

We specialize our general situation

$$\mathrm{Hom}_R(M, W), \quad \mathrm{Hom}_R(W, M)$$

to the pair

$$\mathrm{Hom}_R(R, M), \quad M^* := \mathrm{Hom}_R(M, R).$$

Here M^* is the dual module of M. We further identify $\mathrm{Hom}_R(R, M)$ with M. This is possible since $\mathrm{Hom}_R(R, M)$ and M are isomorphic via

$$\mathrm{Hom}_R(R, M) \ni \beta \mapsto \beta(1) \in M.$$

It is obvious that this is an R–isomorphism. We will see what happens under this isomorphism to a partially invertible $\beta \in \mathrm{Hom}_R(R, M)$. If β is pi, then there exists $\varphi \in M^*$ such that

$$\varphi\beta = e = e^2 \neq 0, \ e \in \mathrm{End}(R_R).$$

Then

$$\varphi\beta(1) = e(1) = e = e^2 \neq 0.$$

Here we have identified the $e \in \mathrm{End}(R_R)$ with left multiplication by e. On the other hand, if there exist $m \in M$, $\varphi \in M^*$ with

$$\varphi(m) = e = e^2 \neq 0,$$

and if $\beta \in \mathrm{Hom}_R(R, M)$ is defined by

$$\beta(x) = mx, \ x \in R,$$

then

$$\varphi\beta(x) = \varphi(mx) = \varphi(m)x = ex,$$

hence

$$\varphi\beta = e, \quad \beta(1) = m.$$

It now makes sense to define pi for elements of M. We also repeat the definition of regularity.

Definition 4.1.

1) $m \in M$ is called **partially invertible** (= pi) if and only if there exists $\varphi \in M^*$ such that
$$e := \varphi(m) = e^2 \neq 0.$$

2) $\mathrm{Tot}(M) := \{m \in M \mid m \text{ is not pi}\}.$

3) $m \in M$ is **regular** if and only if there exists $\varphi \in M^*$ such that

$$m\varphi(m) = m.$$

Other properties connected with pi and the total are similar and easy to transfer from $\mathrm{Hom}_R(R, M)$ to M.

We again set $T := \mathrm{End}(M)$. If

$$e = \varphi(m) = e^2 \neq 0,$$

then, as in the general case, we also have

$$(me\varphi)(me\varphi) = me\varphi(m)e\varphi = me^3\varphi = me\varphi,$$

so that

$$d := me\varphi$$

is an idempotent in T. Since

$$\varphi dm = \varphi me\varphi m = \varphi(m)e\varphi(m) = e^3 = e \neq 0,$$

also $d \neq 0$.

To get results we first consider relations between the totals of R, M, T and M^*. We assume that M is projective. Then we can work with a dual basis of M. Having a dual basis means: For every family $\{y_i \mid i \in I\}$ of generators of M, there exists a family $\{\varphi_i \mid i \in I\}$ of elements $\varphi_i \in M^*$ with

(a) $\forall x \in M, \varphi_i(x) \neq 0$ only for finitely many $i \in I$];

(b) $\forall x \in M, x = \sum_{\varphi_i(x) \neq 0} y_i \varphi_i(x)$.

(See [15, p. 120], the empty summation is 0)

We know that the radical is contained in the total. This property did not get lost in our identifications. If $m \in \mathrm{Rad}(M)$, then for $\varphi \in M^*$, again $\varphi(m) \in \mathrm{Rad}(R)$, hence $\varphi(m)$ cannot be a non–zero idempotent, hence $m \in \mathrm{Tot}(M)$.

We now consider situations in which the radical is equal to the total.

Proposition 4.2. *Assume that $M \in$ Mod–R is projective. Set $T = \mathrm{End}_R(M)$.*

1) *If* $\mathrm{Rad}(R) = \mathrm{Tot}(R)$, *then* $\mathrm{Rad}(M) = \mathrm{Tot}(M)$.

2) *If M is finitely generated and if* $\mathrm{Rad}(M) = \mathrm{Tot}(M)$, *then* $\mathrm{Rad}(T) = \mathrm{Tot}(T)$.

3) *If* $\mathrm{Rad}(T) = \mathrm{Tot}(T)$, *then* $\mathrm{Rad}(M) = \mathrm{Tot}(M)$.

Proof. 1) Since always $\mathrm{Rad}(M) \subseteq \mathrm{Tot}(M)$, we have only to show that $\mathrm{Tot}(M) \subseteq \mathrm{Rad}(M)$. Let $x \in M$. Then

$$x = \textstyle\sum_{\text{finite}} y_i \varphi_i(x)$$

utilizing a dual basis. Suppose further that $x \in \mathrm{Tot}(M)$. Then $\varphi_i(x) \in \mathrm{Tot}(R) = \mathrm{Rad}(R)$. Since $M\,\mathrm{Rad}(R) \subseteq \mathrm{Rad}(M)$ (see [15, p. 218]), we have $y_i\varphi_i(x) \in \mathrm{Rad}(M)$, hence $x \in \mathrm{Rad}(M)$.

2) Again, only $\mathrm{Tot}(T) \subseteq \mathrm{Rad}(T)$ needs to be shown. For $t \in \mathrm{Tot}(T)$ it follows that $t(M) \subseteq \mathrm{Tot}(M)$. Since $\mathrm{Tot}(M) = \mathrm{Rad}(M)$, also $t(M) \subseteq \mathrm{Rad}(M)$. Since M is finitely generated, $\mathrm{Rad}(M) \subseteq^\circ M$ ([15, p. 218]). Therefore $t(M) \subseteq^\circ M$. Thus $t \in \mathrm{Rad}(T)$ (see [15, p. 231]).

3) Again, $\mathrm{Tot}(M) \subseteq \mathrm{Rad}(M)$ needs to be proved. Assume that $m \in \mathrm{Tot}(M)$, and let U be a submodule of M such that

$$M = mR + U.$$

If $\{u_i \mid i \in I\}$ is a family of generators of U, then $\{m, u_i \mid i \in I\}$ is a family of generators of M. There exists a family $\{\varphi, \varphi_i \mid i \in I\}$, $\varphi, \varphi_i \in M^*$, such that

$$\{m, u_i \mid i \in I\}, \quad \{\varphi, \varphi_i \mid i \in I\}$$

are dual bases. Since $m \in \mathrm{Tot}(M)$, it follows that $m\varphi \in \mathrm{Tot}(T) = \mathrm{Rad}(T)$. Since M is projective, $\mathrm{Im}(m\varphi) = m\varphi(M)$ is small in M (see [15, p. 231]). Now it follows that

$$M = m\varphi(M) + \textstyle\sum_{i\in I} u_i\varphi_i(M) = \sum_{i\in I} u_i\varphi_i(M) \subseteq U,$$

hence $U = M$. This implies that $mR \subseteq^\circ M$, so $m \in \mathrm{Rad}(M)$. □

We believe that Lemma 3.1 and Theorem 3.2 are noteworthy also in the special case of a pair M, M^*. Since there is a change of notation, for the convenience of the reader, we restate Theorem 3.2 in this special case.

Special Case of Theorem 3.2. *For an arbitrary module M, one of the following statements is true.*

1) $M = \mathrm{Tot}(M)$.
2) $M = U \oplus \bigoplus_{i=1}^{n} m_i R$ *with* $n \geq 1$, $U \subseteq \mathrm{Tot}(M)$, m_i *regular, and* $m_i R$ *projective for* $i = 1, \ldots, n$.
3) *M contains a finitely direct summand of the form $\bigoplus_{i\in\mathbb{N}} m_i R$ where the $m_i R$ have the same properties as in 2).*

An example for 1) is $\mathbb{Q}_\mathbb{Z}$. It is well–known that $\mathrm{Rad}(\mathbb{Q}_\mathbb{Z}) = \mathbb{Q}_\mathbb{Z}$ (see [15, p. 217]), and since $\mathrm{Rad}(\mathbb{Q}_\mathbb{Z}) \subseteq \mathrm{Tot}(\mathbb{Q}_\mathbb{Z})$, also $\mathrm{Tot}(\mathbb{Q}_\mathbb{Z}) = \mathbb{Q}_\mathbb{Z}$. This also follows from the fact that the dual module $\mathbb{Q}_\mathbb{Z}^*$ is 0.

If A is a non–zero projective module, then $\mathrm{Rad}(A) \neq A$ (see [15, p. 233]). Is the same true for the total? The answer is no! We give an example of a projective module $A \neq 0$ with $\mathrm{Tot}(A) = A$. This is then also an example for $\mathrm{Rad}(A) \subsetneqq \mathrm{Tot}(A)$. We precede the example with a remark.

Remark. In a commutative ring T without zero divisors, any ideal is directly indecomposable.

Proof. Assume that the non–zero ideal A has a decomposition

$$A = A_1 \oplus A_2, \quad A_1 \neq 0.$$

For $a_1 \in A_1$, $a_1 \neq 0$, and $a_2 \in A_2$, it follows that $a_1a_2 = a_2a_1 \in A_1 \cap A_2 = 0$. Hence $a_2 = 0$, so $A_2 = 0$. □

Example 4.3. In the ring $T := \mathbb{Z}[\sqrt{-5}]$ the ideal $A = T3 + T(1 + \sqrt{-5})$ is an example of a projective module A with $\operatorname{Tot}(A) = A$ while $\operatorname{Rad}(A) = 0 \neq \operatorname{Tot}(A)$.

Proof. The ring $T := \mathbb{Z}[\sqrt{-5}]$ is a subring of the field of complex numbers, hence a commutative domain without zero divisors. We apply the norm of complex numbers to T. If $a + b\sqrt{-5} \in T$, $a, b \in \mathbb{Z}$, then $\mathrm{N}(a + b\sqrt{-5}) = a^2 + 5b^2$. Let A be the ideal of T generated by 3 and $1 + \sqrt{-5}$. The elements of A have the form

$$\begin{aligned} \alpha &= 3(a_1 + a_2\sqrt{-5}) + (1 + \sqrt{-5})(b_1 + b_2\sqrt{-5}) \\ &= (3a_1 + b_1 - 5b_2) + (3a_2 + b_1 + b_2)\sqrt{-5}, \quad a_1, a_2, b_1, b_2 \in \mathbb{Z}. \end{aligned}$$

Set

$$c_1 := 3a_1 + b_1 - 5b_2, \quad c_2 := 3a_2 + b_1 + b_2,$$

then $\mathrm{N}(\alpha) = c_1^2 + 5c_2^2$. We intend to show that

$$\mathrm{N}(\alpha) \geq 5 \text{ for } 0 \neq \alpha \in A.$$

If $c_2 \neq 0$, then $\mathrm{N}(\alpha) \geq 5$. If $c_2 = 0$, then $c_1 \neq 0$, and

$$\begin{aligned} & c_2 = 3a_2 + b_1 + b_2 = 0 \\ \Rightarrow\ & b_1 = -3a_2 - b_2 \\ \Rightarrow\ & c_1 = 3a_1 - 3a_2 - b_2 - 5b_2 = 3(a_1 - a_2 - 2b_2), \end{aligned}$$

hence $\mathrm{N}(\alpha) \geq 9$.

We show next that A is not cyclic.

Proof. Assume $A = \alpha_0 T$. Then there exist $\beta, \gamma \in T$ such that

$$\alpha_0\beta = 3, \quad \alpha_0\gamma = 1 + \sqrt{-5}.$$

These imply that $\mathrm{N}(\alpha_0)\,\mathrm{N}(\beta) = \mathrm{N}(\alpha_0\beta) = \mathrm{N}(3) = 9$, and $\mathrm{N}(\alpha_0)\,\mathrm{N}(\gamma) = \mathrm{N}(\alpha_0\gamma) = \mathrm{N}(-1 + \sqrt{-5}) = 6$. Therefore $\mathrm{N}(\alpha_0)$ is a common divisor of 9 and 6, that is 1 or 3, contradicting the fact that $\mathrm{N}(\alpha_0) \geq 5$. □

Now we apply the special case of Theorem 3.2 to A_T. The case 2) is not possible. In fact, A is directly indecomposable, so $U = 0$ and $A = \alpha_1 R$ which contradicts the fact that A is not cyclic. By the same reasoning 3) is not possible, hence 1), i.e., $A = \operatorname{Tot}(A)$ is satisfied.

We still have to prove that A is projective. This follows from general results but here we give a direct proof by showing that A has a dual basis. Denote by φ_1

the multiplication in A by $\frac{1}{3}(2+\sqrt{-5})$ and by φ_2 the multiplication in A by -1. Then $\varphi_1, \varphi_2 \in A^*$ and for $\alpha \in A$ it follows that

$$(3\varphi_1 + (1+\sqrt{-5})\varphi_2)(\alpha) = (2+\sqrt{-5}-1-\sqrt{-5})\alpha = 1\alpha = \alpha,$$

hence $3\varphi_1 + (1+\sqrt{-5})\varphi_2 = 1_A$, i.e., we have a dual basis and A is projective.

It is easy to see that $\mathrm{Rad}(A) = 0$. The only invertible elements in T are 1 and -1 since the norm of an invertible element must be 1. If we assume that $0 \neq u \in \mathrm{Rad}(T)$, then $1-u$ must be invertible, hence $u = 2$. Since $\mathrm{Rad}(T)$ is an ideal, also $1-2u$ must be invertible resulting in the contradiction $u = 1$. With the inclusion map $\iota : A \to T$ we have

$$\iota(\mathrm{Rad}(A)) \subseteq \mathrm{Rad}(T) = 0,$$

hence $\mathrm{Rad}(A) = 0$ for any ideal A of T. □

We return to the special case of Theorem 3.2. There it is interesting to have modules M with $\mathrm{Tot}(M) = 0$. Then 1) cannot occur, and in 2) necessarily $U = 0$. To provide examples we use results which we prove later.

Remark.

1) If R is an injective ring and $\mathrm{Rad}(M) = 0$, then $\mathrm{Tot}(M) = 0$.

2) If W is projective, semiperfect, and if $\mathrm{Rad}(W) = 0$, then $\mathrm{Tot}(W) = 0$.

Proof. 1) Since injective modules are locally injective (by III.3.1) and large restricted (by III.2.4) it follows for injective R that

$$\mathrm{Rad}(R, M) = \mathrm{Tot}(R, M),$$

hence $\mathrm{Rad}(M) = \mathrm{Tot}(M)$, and for $\mathrm{Rad}(M) = 0$, also $\mathrm{Tot}(M) = 0$.

2) Since projective and semiperfect modules are locally projective (by III.3.1) and projective modules are small restricted (by III.2.4), it follows that

$$\mathrm{Rad}(R, W) = \mathrm{Tot}(R, W),$$

hence $\mathrm{Rad}(W) = \mathrm{Tot}(W)$ and then for $\mathrm{Rad}(W) = 0$ also $\mathrm{Tot}(W) = 0$. □

Also III.2.3 provides for numerous examples where the total is zero.

5 Two Examples: $\mathbb{Z}/n\mathbb{Z}$ and modules of finite length

All the notions and properties for rings that occurred in this chapter will be presented explicitly for the ring $\mathbb{Z}/n\mathbb{Z}$, $2 \leq n \in \mathbb{N}$. The properties of $\mathbb{Z}/n\mathbb{Z}$ are well–known, but we wish to consider these in view of partial invertibility and the total.

Before doing so, we show in a diagram the containments of certain subsets of a ring R. This will help the orientation of the reader.

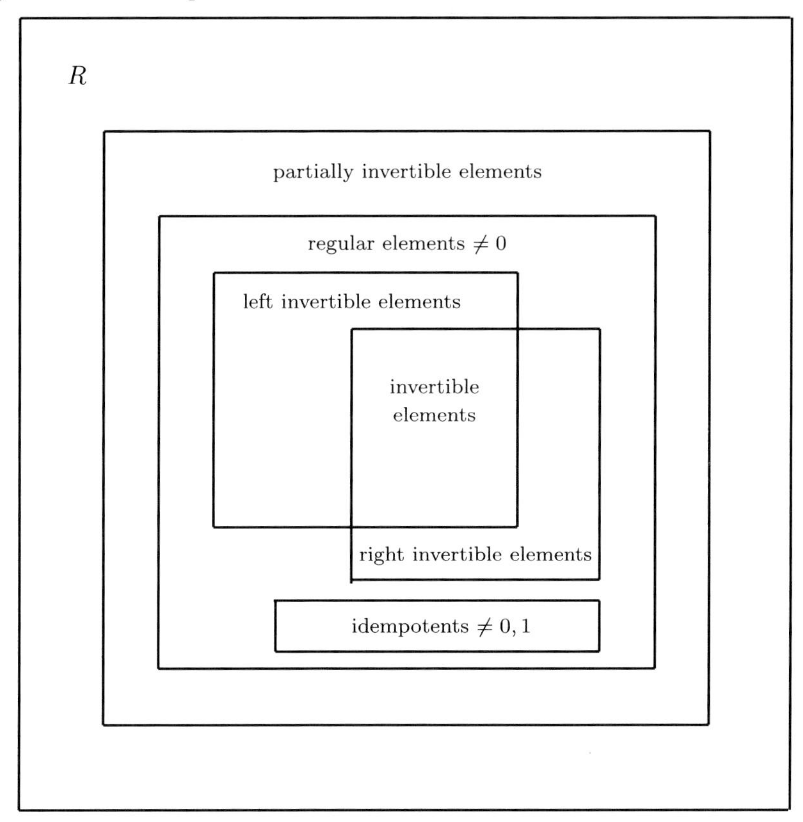

Here we call an element $r \in R$ left– and right–invertible if there exists $s \in R$ with, respectively, $sr = 1$ and $rs = 1$. Either way, it then follows that $rsr = r$, hence r is regular. In a commutative ring left– or right–invertible means invertible.

Now let $2 \leq n \in \mathbb{N}$ with prime factorization

$$n = p_1^{k_1} p_2^{k_2} \cdots p_m^{k_m} \tag{17}$$

where the p_i are different prime numbers and the k_i are positive integers. Set $\overline{z} = z + n\mathbb{Z}$.

If $x, y \in \mathbb{Z}$, then (x, y) denotes the greatest common divisor of x and y. If x is a divisor of y we write $x \mid y$ and otherwise $x \nmid y$.

In order to determine which elements are partially invertible, we need information about idempotents. For this we define

$$a_i := np_i^{-k_i}, \quad i = 1, \ldots m;$$

then the greatest common divisor of $a_1, \dots, a_m$ is 1. Therefore, there exist $b_1, \dots,$ $b_m \in \mathbb{Z}$ such that

$$1 = a_1 b_1 + \cdots + a_m b_m.$$

Set $e_i := a_i b_i$, $i = 1, \dots, m$. Then

$$1 = e_1 + \cdots + e_m \tag{18}$$

and

$$\overline{1} = \overline{e_1} + \cdots + \overline{e_m} \in \mathbb{Z}/n\mathbb{Z}. \tag{19}$$

Proposition 5.1. *The family $\{\overline{e_1}, \dots, \overline{e_m}\}$ is an orthogonal set of non-zero idempotents in $\mathbb{Z}/n\mathbb{Z}$.*

Proof. Orthogonal: By definition of the a_i we have

$$n \mid a_i a_j, \quad i \neq j, \quad i, j \in \{1, \dots, m\}.$$

Hence

$$n \mid e_i e_j,\ i \neq j, \quad \text{therefore} \quad \overline{e_i}\,\overline{e_j} = \overline{0},\ i \neq j.$$

Idempotents: Since the $\overline{e_1}, \dots, \overline{e_m}$ are orthogonal, it follows from (19) that $\overline{e_i} = \overline{e_i}^2$.
Non-zero: By definition of the a_i, $p_i \mid a_j$, $i \neq j$, hence $p_i \mid e_j$, and, by (18),

$$p_i \nmid e_i. \tag{20}$$

Therefore $n \nmid e_i$ and $\overline{e_i} \neq \overline{0}$. □

Now we come to the characterization of the different invertibility properties.

Proposition 5.2. *Let $z \in \mathbb{Z}$ and $\overline{z} \in \mathbb{Z}/n\mathbb{Z}$. Then the following statements hold.*

1) *$\overline{z}$ is partially invertible if and only if there is $i \in \{1, \dots, m\}$ such that $p_i \nmid z$.*
2) *$\overline{z}$ is regular if and only if for all $i \in \{1, \dots, m\}$ either $p_i \nmid z$ or $p_i^{k_i} \mid z$.*
3) *$\overline{z}$ is invertible if and only if $p_i \nmid z$ for all $i \in \{1, \dots, m\}$ if and only if $(z, n) = 1$.*

Proof. 1) $\Leftarrow$: Assume $p_i \nmid z$. Then there exist $x, y \in \mathbb{Z}$ with

$$zx + p_i^{k_i} y = 1. \tag{21}$$

Since for $e_i = a_i b_i$ we have that $n \mid p_i^{k_i} e_i$, it follows that

$$\overline{p_i^{k_i}}\,\overline{e_i} = \overline{0} \in \mathbb{Z}/n\mathbb{Z},$$

and further by (21) that

$$\overline{z}\,\overline{x}\,\overline{e_i} = \overline{e_i},$$

which means that $\overline{z}$ is pi.

1) $\Rightarrow$: Indirect. Assume that $p_i \mid z$ for $i = 1, \ldots, m$ and

$$\overline{z}\,\overline{x} = \overline{d} = \overline{d}^2 \neq \overline{0}.$$

Then $p_1 p_2 \cdots p_m \mid z$. Let $k = \max\{k_1, \ldots, k_m\}$. Then

$$n \mid (p_1 p_2 \cdots p_m)^k \mid (zx)^k$$

and this implies that

$$\overline{0} = (\overline{z}\,\overline{x})^k = \overline{d}^k = \overline{d} \neq \overline{0},$$

a contradiction.

2) $\Leftarrow$: If, for all i, we have $p_i^{k_i} \mid z$, then $n \mid z$, hence $\overline{z} = \overline{0}$ which is regular. Now for an i with $p_i \nmid z$, there exist (again as in (21)) $x_i, y_i \in \mathbb{Z}$ with

$$zx_i + p_i^{k_i} y_i = 1$$

and this implies again

$$\overline{z}\,\overline{x_i}\,\overline{e_i} = \overline{e_i}. \tag{22}$$

If $p_j^{k_j} \mid z$, then it follows by definition of e_j that $n \mid ze_j$, hence

$$\overline{z}\,\overline{e_j} = \overline{0}. \tag{23}$$

Now set

$$c := \sum_{i \in \{1,\ldots,m\}, p_i \nmid z} x_i e_i + \sum_{j \in \{1,\ldots,m\}, p_j^{k_j} \mid z} e_j.$$

For the sake of brevity, we write $\sum_i$ for the summation over $i \in \{1, \ldots, m\}, p_i \nmid z$ and $\sum_j$ for the summation over $j \in \{1, \ldots, m\}, p_j^{k_j} \mid z$. It then follows from (22) and (23) that

$$\begin{aligned} \overline{z}\,\overline{c}\,\overline{z} &= \left(\textstyle\sum_i \overline{z}\,\overline{x_i}\,\overline{e_i} + \sum_j \overline{z}\,\overline{e_j}\right) \overline{z} \\ &= \left(\textstyle\sum_i \overline{e_i} + \overline{0}\right) \overline{z} = \left(\textstyle\sum_i \overline{e_i} + \sum_j \overline{e_j}\right) \overline{z} = \overline{z}. \end{aligned}$$

Here we used the fact that, by (23),

$$\textstyle\sum_j \overline{z}\,\overline{e_j} = \overline{0} = \sum_j \overline{e_j}\,\overline{z}.$$

2) $\Rightarrow$: Indirect. We assume that for $z, x \in \mathbb{Z}$ we have

$$\overline{z}\,\overline{x}\,\overline{z} = \overline{z} \neq \overline{0} \tag{24}$$

and that there exists an $i \in \{1, \ldots, m\}$ such that

$$p_i^\ell \mid z \text{ with } 1 \leq \ell < k_i, \text{ and } p_i^{\ell+1} \nmid z.$$

Write $z = p_i^\ell z_0$. By (24) there exists $y \in \mathbb{Z}$ with

$$zxz - z = ny.$$

Hence

$$zxz - ny = p_i^{2\ell}(z_0 x z_0) - p_i^{k_i} a_i y = z = p_i^\ell z_0. \tag{25}$$

Since $2\ell \geq \ell + 1$ and $k_i \geq \ell + 1$, the left-hand side of (25) is divisible by $p^{\ell+1}$, but the right-hand side, which is $z = p^\ell z_0$ is not, a contradiction.

3) Clear. $\square$

The idempotents are fundamental for the definition of the partially invertible elements. Therefore we ask for all idempotents of $\mathbb{Z}/n\mathbb{Z}$. The answer is that all the partial sums of

$$\overline{1} = \overline{e_1} + \overline{e_2} + \cdots + \overline{e_m}$$

are all the idempotents of $\mathbb{Z}/n\mathbb{Z}$.

To prove this, let d be an integer so that $\overline{d}$ is a non–zero idempotent. Then, by (19),

$$\overline{d} = \overline{e_1}\,\overline{d} + \overline{e_2}\,\overline{d} + \cdots + \overline{e_m}\,\overline{d}$$

and there exists i with $\overline{e_i}\,\overline{d} \neq \overline{0}$. Clearly, $p_i \mid d$ and $p_i \mid 1-d$ are mutually exclusive. Therefore we have to consider two cases.

Case $p_i \nmid d$: There exist $x, y \in \mathbb{Z}$ with

$$dx + p_i^{k_i} y = 1.$$

This implies

$$\overline{e_i}\,\overline{d}\,\overline{x} = \overline{e_i} \quad (n \mid a_i p_i^{k_i} \text{ so } \overline{e_i}\,\overline{p_i}^{k_i}\,\overline{y} = \overline{0}).$$

Multiplying by $\overline{d}$ we find

$$\overline{e_i}\,\overline{d}^2\,\overline{x} = \overline{d}\,\overline{e_i} = \overline{e_i}\,\overline{d}\,\overline{x} = \overline{e_i},$$

hence

$$\overline{d}\,\overline{e_i} = \overline{e_i}.$$

Case $p_i \nmid 1 - d$: It follows analogously to the first case that

$$\overline{e_i} = \overline{e_i}(\overline{1} - \overline{d}) = \overline{e_i} - \overline{e_i}\,\overline{d},$$

hence

$$\overline{e_i}\,\overline{d} = \overline{0}.$$

Both cases together imply that

$$\overline{d} = \sum_{i \in \{1,\ldots,m\},\, \overline{e_i}\,\overline{d} \neq \overline{0}} \overline{e_i}\,\overline{d} = \sum \overline{e_i},$$

hence $\overline{d}$ is a partial sum of (19).

Now we ask for $\mathrm{Tot}(\mathbb{Z}/n\mathbb{Z})$. This is easy to answer with Proposition 5.2.1). Which are the z such that $\overline{z}$ is not pi? These are the z such that every p_i divides z, hence $p_1p_2\cdots p_m \mid z$:

$$\mathrm{Tot}(\mathbb{Z}/nZ) = \overline{p_1p_2\cdots p_m}\ \mathbb{Z}/n\mathbb{Z}.$$

But this is also the radical of $\mathbb{Z}/n\mathbb{Z}$ (see [15, p. 217]). Here we have an example of a ring whose radical is equal to the total. Note also that $\mathrm{Tot}(\mathbb{Z}_n) = 0$ if and only if n is square–free.

As another interesting example we consider now a module $M \neq 0$ of finite length. We present explicitly the total of the endomorphism ring of M, $\mathrm{Tot}(\mathrm{End}(M))$, and produce all idempotents in $\mathrm{End}(M)$, needed to determine the partially invertible elements of M.

A module M of finite length has the following properties.

i) M has a finite decomposition $M = M_1 \oplus \cdots \oplus M_n$ with indecomposable summands $M_i \neq 0$ that have local endomorphism rings $\mathrm{End}(M_i)$, $i = 1, \ldots, n$.

ii) If also $M = N_1 \oplus \cdots \oplus N_k$ is a decomposition with non–zero indecomposable direct summands N_j, then $n = k$ and there exists an automorphism $\alpha \in \mathrm{End}(M)$ and a permutation ρ of the set $\{1, \ldots n\}$ such that

$$M_{\rho(i)} \ni x \mapsto \alpha(x) \in N_i, \quad i = 1, \ldots, n,$$

is an isomorphism.

iii) Any direct decomposition of M can be refined to a composition as in ii).

These are well–known facts. For the existence of a decomposition as in i) see, e.g., [15, 7.2.9] and the uniqueness of the decomposition up to isomorphisms is proved by the theorem of Remak–Krull–Schmidt ([15, 7.2.9]). Finally, iii) follows from the fact that a direct summand of a module of finite length has again finite length and therefore has a direct decomposition with finitely many indecomposable summands.

We set $I_n := \{1, \ldots, n\}$, and consider subsets $\emptyset \neq J \subseteq I_n$. Then let e_J be the idempotent of $\mathrm{End}(M)$ with

$$e_J(M) = \bigoplus_{j \in J} M_j, \quad e_J\Big(\bigoplus_{i \in I_n \setminus J} M_i\Big) = 0.$$

Proposition 5.3. *Let M be a non–zero module of finite length. Then the following statements hold.*

1) *Any idempotent $0 \neq e \in \mathrm{End}(M)$ is of the form $\alpha e_J \alpha^{-1}$ for some $\emptyset \neq J \subseteq I_n$ and some $\alpha \in \mathrm{Aut}(M)$.*

2) *Let*

$$F := \{f \in \mathrm{End}(M) \mid \exists g \in \mathrm{End}(M), \exists \emptyset \neq J \subseteq I_n : gf = e_J\}.$$

Then

$$\mathrm{Tot}(\mathrm{End}(M)) = \mathrm{End}(M) \setminus \bigcup\{\alpha F \alpha^{-1} \mid \alpha \in \mathrm{Aut}(M)\}.$$

Proof. 1) Let $0 \neq e = e^2 \in \mathrm{End}(M)$. Then $M = e(M) \oplus (1-e)(M)$. By refinement a decomposition as in ii) is obtained:

$$e(M) = \bigoplus_{i=1}^{m} N_i, \quad (1-e)(M) = \bigoplus_{i=m+1}^{n} N_i.$$

(If $e = 1$, then $m = n$) Let, in the sense of ii), $J := \{\rho(i) \mid i = 1, \ldots, m\}$. Then we have the isomorphisms

$$M_{\rho(i)} \ni x \mapsto \alpha(x) \in N_i, \quad i \in I_n,$$

and by applying $\alpha e_J \alpha^{-1}$ to the decomposition $M = \bigoplus_{i \in I_n} N_i$, we see that $e = \alpha e_J \alpha^{-1}$.

2) Let $0 \neq e = e^2 \in \mathrm{End}(M)$ and $\alpha \in \mathrm{Aut}(M)$. Then $(\alpha e \alpha^{-1})(\alpha e \alpha^{-1}) = \alpha e \alpha^{-1}$ is an idempotent. Suppose further that k is partially invertible, say $kh = e = e^2 \neq 0$, $e, k, h \in \mathrm{End}(M)$. Then

$$\alpha e \alpha^{-1} = (\alpha k \alpha^{-1})(\alpha h \alpha^{-1})$$

and $\alpha k \alpha^{-1}$ is also partially invertible. In particular, $\alpha F \alpha^{-1}$ contains only partially invertible elements. On the other hand, suppose that $0 \neq kh = e = e^2$. Then, by 1), $e = \alpha e_J \alpha^{-1} = kh$, and this implies that $\alpha^{-1} e \alpha = e_J = (\alpha^{-1} k \alpha)(\alpha^{-1} h \alpha)$, hence $\alpha^{-1} h \alpha \in F$, and then $h \in \alpha F \alpha^{-1}$. Together it follows that the set $\bigcup\{\alpha F \alpha^{-1} \mid \alpha \in \mathrm{Aut}(M)\}$ is the set of all pi elements in $\mathrm{End}(M)$, hence $\mathrm{Tot}(\mathrm{End}(M))$ is the complementary set in $\mathrm{End}(M)$. □

6 Appendix on semi–ideals

The total is a semi–ideal, and it is therefore desirable to gain insight into the algebraic nature of semi–ideals. The following proposition states more generally what was noted in Proposition 1.11. Recall that $\mathrm{Hom}_R(M, W)$ is an $\mathrm{End}(W)$–$\mathrm{End}(M)$–bimodule.

Proposition 6.1. *Let $M, W \in$ Mod–R and let T be a semi–ideal contained in $\mathrm{Hom}_R(M, W)$. Then T is the set–theoretic union of cyclic left $\mathrm{End}(W)$–submodules and also the set–theoretic union of cyclic right $\mathrm{End}(M)$–submodules.*

In order to further clarify the structure of semi–ideals one might hope that a semi–ideal, such as the total, is the union of cyclic submodules in which unnecessary submodules are omitted in the following precise sense.

Definition 6.2. Let $\mathcal{F}$ be a family of subsets of a set U.

1) The family $\mathcal{F}$ is an **irredundant family** if for every $F \in \mathcal{F}$ it is the case that $F \not\subseteq \bigcup \mathcal{F}$.
2) The family $\mathcal{F}$ **covers** $V \subseteq U$ if $V \subseteq \bigcup \mathcal{F}$.

The following example shows that it is not always possible to replace a family of cyclic modules that covers the total, a specific semi–ideal, by an irredundant family that still covers the total.

Example 6.3. Let $\mathbb{Z}(p^\infty)$ be the Prüfer group. Then

$$\operatorname{Tot}(\mathbb{Z}, \mathbb{Z}(p^\infty)) = \operatorname{Hom}(\mathbb{Z}, \mathbb{Z}(p^\infty)) \cong \mathbb{Z}(p^\infty),$$

and there is no irredundant family of cyclic subgroups of $\mathbb{Z}(p^\infty)$ that covers $\mathbb{Z}(p^\infty)$.

Proof. The Prüfer group can be defined as follows:

$$\mathbb{Z}(p^\infty) := A/\mathbb{Z} \quad \text{where} \quad A := \{m/p^k \mid m \in \mathbb{Z}, k \in \mathbb{N}_0\} \subseteq \mathbb{Q}. \tag{26}$$

It can be verified that $\langle 1/p^i + \mathbb{Z}\rangle$ is a (cyclic) subgroup of order p_i, $\langle 1/p^i + \mathbb{Z}\rangle \subseteq \langle 1/p^{i+1} + \mathbb{Z}\rangle$ and $\mathbb{Z}(p^\infty) = \bigcup_{i=1}^{\infty} \langle 1/p^i + \mathbb{Z}\rangle$. Clearly, in any cover of $\mathbb{Z}(p^\infty)$ by cyclic subgroups there will be redundant ones. We remark that $\mathbb{Z}(p^\infty)$ is a divisible p–primary group, $\operatorname{End}(\mathbb{Z}(p^\infty)) = \hat{\mathbb{Z}}_p$, the ring of p–adic integers, and the $\hat{\mathbb{Z}}_p$–submodules of $\mathbb{Z}(p^\infty)$ are exactly the subgroups of $\mathbb{Z}(p^\infty)$ (see [11]). □

It is clear now that irredundant families of cyclic submodules can only be expected if every cyclic submodule is contained in a maximal one. This necessary condition turns out to be sufficient.

Proposition 6.4. *Let T be a semi–ideal in $\operatorname{Hom}_R(M, W)$. Assume that every cyclic submodule of T is contained in a maximal cyclic submodule of T. Then T is the union of an irredundant family of cyclic submodules.*

Proof. By Proposition 6.1 there exists a family of cyclic submodules of T that covers T. Let $\mathcal{F}$ be a family of cyclic submodules of T that covers T. By hypothesis every cyclic submodule in $\mathcal{F}$ can be replaced by a maximal one. Therefore we assume without loss of generality that every $F \in \mathcal{F}$ is a maximal cyclic submodule of T.

Let $\mathcal{S}$ be the family of all irredundant subfamilies of $\mathcal{F}$. Then $\mathcal{S}$ is partially ordered by inclusion and is non–void since singletons of elements in $\mathcal{F}$ are irredundant. Let $\mathcal{C}$ be a chain in $\mathcal{S}$. We wish to show that $\bigcup \mathcal{C} \in \mathcal{S}$, i.e., we wish to show that $\bigcup \mathcal{C}$ is irredundant. Let $S = \operatorname{End}_R(M)$ and let us specifically assume that we are talking about right S–modules. The story for left $\operatorname{End}_R(W)$–modules will be analogous. Then we wish to show that $xS \in \mathcal{C}$ implies that $xS \not\subseteq \bigcup(\mathcal{C} \setminus \{xS\})$. Assume to the contrary that $xS \in \mathcal{C}$ and $xS \subseteq \bigcup(\mathcal{C} \setminus \{xS\})$. Then there is $C \in \mathcal{C}$ and $xS \neq yS \in C$ such that $x \in yS$ and hence $xS \subseteq yS$. But xS is maximal cyclic and we obtain the contradiction that $xS = yS$. Thus $\bigcup \mathcal{C}$ is irredundant and by Zorn's Lemma there are maximal irredundant subfamilies of $\mathcal{F}$. Let $\mathcal{M}$ be a maximal irredundant subfamily of $\mathcal{F}$. We claim that $\mathcal{M}$ covers T. Let $x \in T$. Then $x \in zS$ for some cyclic submodule $zS \in \mathcal{F}$. If $zS \subseteq \bigcup \mathcal{M}$, then $x \in \bigcup \mathcal{M}$ and we are done. Suppose to the contrary that $zS \not\subseteq \bigcup \mathcal{M}$. Then $z \notin \bigcup \mathcal{M}$ and $zS \notin \mathcal{M}$. Consider now $\mathcal{M}' := \mathcal{M} \cup \{zS\}$. Then $\mathcal{M}'$ is a subfamily of $\mathcal{F}$, and if we show that it is irredundant, then we have a contradiction to the maximality of $\mathcal{M}$ and can conclude that $\mathcal{M}$ does indeed cover T as desired.

To show that $\mathcal{M}'$ is irredundant first note that $zS \not\subseteq \bigcup(\mathcal{M}' \setminus \{zS\}) = \bigcup \mathcal{M}$. Next let $zS \neq yS \in \mathcal{M}'$ and assume that $yS \subseteq \bigcup(\mathcal{M}' \setminus \{yS\})$. Then $y \in \bigcup(\mathcal{M}' \setminus \{yS\}) = \bigcup((\mathcal{M} \cup \{zS\}) \setminus \{yS\})$. Now $y \notin zS$, else $yS \subseteq zS$ and by maximality $yS = zS$ contrary to assumption. So $y \in \bigcup(\mathcal{M} \setminus \{yS\})$, so $yS \subset \bigcup(\mathcal{M} \setminus \{yS\})$ contradicting the fact that $\mathcal{M}$ was irredundant. $\square$

Chapter III

Good conditions for the total

Here the question is: How do the properties of $\mathrm{Tot}(M, W)$ depend on "good conditions" for M or W? In particular, we are interested in getting results concerning the question: When are $\mathrm{Rad}(M, W)$, $\Delta(M, W)$ and $\nabla(M, W)$ equal to $\mathrm{Tot}(M, W)$? If this is the case, then $\mathrm{Tot}(M, W)$ is also additively closed.

Good tools for answering these questions are four notions that are pairwise dual to each other. The first two are restrictions on certain large or small submodules (LR–modules or SR–modules). The others are called locally injective and locally projective. We have the best conditions if both occur simultaneously. They are both satisfied if either the module is injective or else projective and semiperfect.

1 Restricted modules

Definition 1.1.

1) The module V is called **restricted for large submodules** or an LR–**module** if and only if every monomorphism $f : V \to V$ with $\mathrm{Im}(f) \subseteq^* V$ is an automorphism, i.e., $\mathrm{Im}(f) = V$.

2) The module W is called **restricted for small submodules** or an SR–**module** if and only if every epimorphism $f : W \to W$ with $\mathrm{Ker}(f) \subseteq^\circ W$ is an automorphism, i.e., $\mathrm{Ker}(f) = 0$.

Since every monomorphism of an injective module and every epimorphism on a projective module splits, injective modules are LR and projective modules are SR.

Theorem 1.2.

1) *If V is an* LR*–module, then for every $M \in$ Mod–R,*

$$\Delta(V, M) \subseteq \mathrm{Rad}(V, M).$$

2) *If* W *is an* SR–*module, then for every* $M \in \text{Mod–}R$,

$$\nabla(M, W) \subseteq \text{Rad}(M, W).$$

Proof. 1) If $f \in \Delta(V, M)$ and $g \in \text{Hom}_R(M, V)$, then also $\text{Ker}(gf) \subseteq^* V$ and this implies that

$$\text{Ker}(gf) \subseteq \text{Im}(1 - gf) \subseteq^* V.$$

Since

$$\text{Ker}(gf) \cap \text{Ker}(1 - gf) = 0$$

and $\text{Ker}(gf) \subseteq^* V$, it follows that $\text{Ker}(1-gf) = 0$. Hence $1-gf$ is a monomorphism of V to V with large image. Since V is an LR–module, the map $1 - gf$ is an automorphism. By Lemma II.2.1 it follows that $f \in \text{Rad}(V, M)$.

2) Assume now that $f \in \nabla(M, W)$ and $g \in \text{Hom}_R(W, M)$. Then also $\text{Im}(fg) \subseteq^\circ W$. Since also

$$\text{Im}(1 - fg) + \text{Im}(fg) = W,$$

we get $\text{Im}(1 - fg) = W$, hence $1 - fg$ is an epimorphism. If $w \in \text{Ker}(1 - fg)$, that is, $w = fg(w)$, then $w \in \text{Im}(fg) \subseteq^\circ W$. Since W is an SR–module, it follows that $1 - fg$ is an automorphism. Again it follows that $f \in \text{Rad}(M, W)$. □

Corollary 1.3.

1) *If* V *is an* LR–*module, then for all* $M \in \text{Mod–}R$,

$$\Delta(V, M) + \text{Tot}(V, M) = \text{Tot}(V, M).$$

2) *If* W *is an* SR–*module, then for all* $M \in \text{Mod–}R$,

$$\nabla(M, W) + \text{Tot}(M, W) = \text{Tot}(M, W).$$

Proof. This follows from Theorem II.2.4 and Theorem 1.2. □

2 Locally injective and locally projective modules

The notions "locally injective module" and "locally projective module" were suggested by the following Theorem 2.2 which also shows immediately that these are worthwhile notions.

Definition 2.1.

1) A module V is called **locally injective** if and only if for every submodule $A \subseteq V$ that is not large (= essential) in V, there exists a non–zero injective submodule Q of V with $A \cap Q = 0$.

2) A module W is called **locally projective** if and only if for every submodule $B \subseteq W$ that is not small in W, there exists a non–zero projective direct summand P of W with $P \subseteq B$.

Theorem 2.2.

1) *For a module V the following properties are equivalent.*

   i) *For every $M \in$ Mod–R, $\Delta(V, M) = \mathrm{Tot}(V, M)$.*

   ii) *V is locally injective.*

2) *For a module W the following properties are equivalent.*

   i) *For every $M \in$ Mod–R, $\nabla(M, W) = \mathrm{Tot}(M.W)$.*

   ii) *W is locally projective.*

Proof. 1) i) $\Rightarrow$ ii): Assume $A \subseteq V$ and A is not large in V. Denote by $\nu : V \to V/A$ the natural epimorphism with $\mathrm{Ker}(\nu) = A$. Let $g : V/A \to D$ be a monomorphism into an injective module D (for example, let D be the injective hull of V/A). Since $\mathrm{Ker}(g\nu) = A$ is not large in V, $g\nu$ is pi by assumption. Then there exist $0 \neq Q \subseteq^{\oplus} V$, $C \subseteq^{\oplus} D$ such that

$$h : Q \ni x \mapsto g\nu(x) \in C$$

is an isomorphism. Since D is injective, also C and Q are injective. Since h is an isomorphism and $\mathrm{Ker}(g\nu) = A$, it follows that $A \cap Q = 0$. Hence V is locally injective.

1) ii) $\Rightarrow$ i): Since we saw already that $\Delta(V, M) \subseteq \mathrm{Tot}(V, M)$, we only have to prove $\mathrm{Tot}(V, M) \subseteq \Delta(V, M)$. To do so we show: If $f \in \mathrm{Hom}_R(V, M)$ and $\mathrm{Ker}(f)$ is not large in V, then f is pi. If $\mathrm{Ker}(f)$ is not large in V, then there exists a non–zero injective $Q \subseteq V$ with $\mathrm{Ker}(f) \cap Q = 0$. Therefore the mapping

$$Q \ni x \mapsto f(x) \in f(Q)$$

is an isomorphism. Since Q and also $f(Q)$ are injective, both are direct summands. This implies that f is pi.

2) i) $\Rightarrow$ ii): Assume that $B \subseteq W$ and B is not small in W. Denote by $\iota : B \to W$ the inclusion and by $g : N \to B$ an epimorphism with projective N (for example take N free). Then $\iota g : N \to W$ has the image B which is not small in W. Hence by assumption ιg is pi and then there exist direct summands $0 \neq K \subseteq^{\oplus} N$, $P \subseteq^{\oplus} W$ such that

$$K \ni x \mapsto \iota g(x) \in P$$

is an isomorphism. Since N is projective, also K and so P are projective. Since $\mathrm{Im}(g) = B$, also $\iota g(K) = P \subseteq B$.

2) ii) $\Rightarrow$ i): Assume $f \in \mathrm{Hom}_R(M, W)$ and $\mathrm{Im}(f)$ is not small in W. Let $0 \neq P \subseteq^{\oplus} W$, P projective, and $P \subseteq \mathrm{Im}(f)$. Since $P \subseteq^{\oplus} W$, there exists a projection $\pi : W \to P$. Since $P \subseteq \mathrm{Im}(f)$, the mapping πf is an epimorphism. Since P is projective, πf splits:

$$M = M_1 \oplus \mathrm{Ker}(\pi f),$$

and then

$$M_1 \ni x \mapsto \pi f(x) \in P$$

is an isomorphism. This means that πf is pi and then also f is pi. □

Corollary 2.3.

1) *For a module V_0 the following statements are equivalent.*
   i) *For all $M \in$ Mod-R, $\mathrm{Tot}(V_0, M) = 0$;*
   ii) *V_0 is a locally injective module and $\mathrm{Soc}(V_0) = V_0$;*
   iii) *V_0 is semisimple and all of its simple submodules are injective.*
2) *For a module W_0 the following statements are equivalent.*
   i) *For all $M \in$ Mod-R, $\mathrm{Tot}(M, W_0) = 0$;*
   ii) *W_0 is a locally projective module and $\mathrm{Rad}(W_0) = 0$;*
   iii) *every non-zero submodule of W_0 contains a non-zero projective direct summand of W_0.*

Proof. 1) i) $\Rightarrow$ ii): Since always $\Delta \subseteq \mathrm{Tot}$, $\mathrm{Tot}(V_0, M) = 0$ implies that $\Delta(V_0, M) = \mathrm{Tot}(V_0, M) = 0$. Hence by Theorem 2.2 it follows that V_0 is locally injective. If V_0 would have a proper large submodule C, then we would have $V_0/C \neq 0$, and

$$(\nu : V_0 \to V_0/C) \in \Delta(V_0, V_0/C),$$

hence $\Delta(V_0, V_0/C) \neq 0$, a contradiction. Hence V_0 is locally injective and does not have a proper large submodule. This implies that $\mathrm{Soc}(V_0) = V_0$ (Theorem 0.10.1)).

1) ii) $\Rightarrow$ i): Since V_0 is locally injective, it follows by Theorem 2.2 that

$$\Delta(V_0, M) = \mathrm{Tot}(V_0, M).$$

Since V_0 does not have a proper large submodule, we have $\Delta(V_0, M) = 0$, hence $\mathrm{Tot}(V_0, M) = 0$.

1) ii) $\Rightarrow$ iii): Since $\mathrm{Soc}(V_0) = V_0$ and $\mathrm{Soc}(V_0)$ is semisimple, also V_0 is semisimple. If B is a simple submodule of V_0 and $V_0 = A \oplus B$, then, A not being large in V_0, there exists an injective submodule $Q \neq 0$ with $A \cap Q = 0$. It then follows (A is maximal since B is simple) that $V_0 = A \oplus Q$. But then $B \cong V_0/A \cong Q$ which implies that B is also injective.

1) iii) $\Rightarrow$ ii): If A is a proper submodule of V_0, then A is a direct summand of V_0, i.e., $V_0 = A \oplus B$, $B \neq 0$. Then B is a direct sum of simple submodules that are all injective. This implies that V_0 is locally injective.

2) i) $\Rightarrow$ ii): By (7) $\nabla(M, W_0) \subseteq \mathrm{Tot}(M, W_0)$ and $\mathrm{Tot}(\mathrm{Rad}(W_0), W_0) = 0$, so $\nabla(M, W_0) = \mathrm{Tot}(M, W_0)$ and by Theorem 2.2.2) W_0 is locally projective. As the inclusion mapping $\iota : \mathrm{Rad}(W_0) \to W_0$ is in $\mathrm{Tot}(\mathrm{Rad}(W_0), W_0) = 0$, we have $\iota = 0$, i.e., $\mathrm{Rad}(W_0) = 0$.

2) ii) $\Rightarrow$ i): By Theorem 2.2.2) it follows that $\nabla(M, W_0) = \mathrm{Tot}(M, W_0)$. Since $\mathrm{Rad}(W_0) = 0$, we have $\nabla(M, W_0) = 0$, therefore $\mathrm{Tot}(M, W_0) = 0$.

2) ii) $\Leftrightarrow$ iii): Clear. □

The results in Corollary 2.3 were first proved in the paper [5] without using the notions of locally injective and locally projective. There the modules V_0 and W_0 were called left– and right–TOTO–modules respectively. (The second O in TOTO stands for zero)

Remark. The question arises whether in Corollary 2.3 also W_0 is semisimple as was V_0. This is the case if we assume that for $A \subseteq W_0$, there exists a submodule B minimal with respect to $A + B = W_0$. Then $A \cap B \subseteq^\circ W_0$ ([15, 5.2.4(a)]) and since $\mathrm{Rad}(W_0) = 0$ it follows that $A \oplus B = W_0$ showing that W_0 is semisimple. However, the existence of such an "addition complement" B of A cannot be asserted.

Corollary 2.4.

1) *If V is locally injective, then $\nabla(V, M) \subseteq \Delta(V, M)$ for all $M \in$ Mod–R.*

2) *If W is locally projective, then $\Delta(M, W) \subseteq \nabla(M, W)$ for all $M \in$ Mod–R.*

Proof. 1) Without assumptions we have $\nabla(V, M) \subseteq \mathrm{Tot}(V, M)$. By Theorem 2.2 our assertion follows.

2) Dual. □

Corollary 2.5.

1) *If V is locally injective and* LR, *then $\Delta(V, M) = \mathrm{Rad}(V, M) = \mathrm{Tot}(V, M)$ for all $M \in$ Mod–R.*

2) *If W is locally projective and* SR, *then $\nabla(M, W) = \mathrm{Rad}(M, W) = \mathrm{Tot}(M, W)$ for all $M \in$ Mod–R.*

Proof. 1) Without assumption we have $\mathrm{Rad}(V, M) \subseteq \mathrm{Tot}(V, M)$. Equality follows from

$$\Delta(V, M) \overset{1.2.1.}{\subseteq} \mathrm{Rad}(V, M) \subseteq \mathrm{Tot}(V, M) \overset{2.2.1.}{=} \Delta(V, M).$$

2) Dual. □

Both conditions on V in Corollary 2.5 are satisfied if V is injective (see Proposition 3.1). Both conditions on W in Corollary 2.5 are satisfied if W is projective and semiperfect (see Proposition 3.1).

An interesting special case in this connection is the total of a module M. In Chapter I, Section 4 we defined $\operatorname{Tot}(M)$ by means of the identification

$$i : \operatorname{Hom}_R(R, M) \ni \beta \mapsto \beta(1) \in M.$$

It is easy to see that this is an isomorphism. The Abelian group $\operatorname{Hom}_R(R, M)$ is turned into a right R–module by the following definition:

$$(\beta r)(x) := \beta(rx), \quad \beta \in \operatorname{Hom}_R(R, M), \ r, x \in R,$$

and i is then an R–isomorphism:

$$i(\beta r) = (\beta r)(1) = \beta(r) = \beta(1)r = i(\beta)r.$$

In the following we apply i without citation.

For $m \in M$ we denote by $\operatorname{Ann}_R(m) := \{r \in R \mid mr = 0\}$ the **annihilator of m in R.** Then we obtain the following corollary as a special case of Theorem 2.2.

Corollary 2.6.

1) *If R_R is locally injective, then*

$$i(\Delta(R, M)) = \{m \in M \mid \operatorname{Ann}_R(m) \subseteq^* R_R\} = i(\operatorname{Tot}(R, M)) = \operatorname{Tot}(M).$$

2) *If R_R is locally projective, then*

$$\nabla(M, R)) = \{\varphi \in M^* \mid \varphi(M) \subseteq \operatorname{Rad}(R)\} = \operatorname{Tot}(M, R)) = \operatorname{Tot}(M^*).$$

Since $\Delta(R, M)$ and $\nabla(M, R)$ are additively closed, also $\operatorname{Tot}(M)$ and $\operatorname{Tot}(M^*)$ are additively closed.

For some module N let $f \in \operatorname{Hom}_R(M, N)$. Then we know by Corollary II.1.10

$$f \operatorname{Tot}(R, M) \subseteq \operatorname{Tot}(R, N).$$

If $\beta \in \operatorname{Tot}(R, M)$, hence $\beta(1) \in \operatorname{Tot}(M)$, then $(f\beta)(1) = f(\beta(1)) \in \operatorname{Tot}(N)$. This means $f(\operatorname{Tot}(M)) \subseteq \operatorname{Tot}(N)$ or in other words: If R_R is locally injective, then $\operatorname{Tot}(M)$ is a **functorial submodule**, i.e., Tot is a functor on Mod–R that assigns to every module $M \in$ Mod–R a submodule $\operatorname{Tot}(M)$ and to every map $f : M \to N$ the map $\operatorname{Tot}(f) := f \restriction_{\operatorname{Tot}(M)}: \operatorname{Tot}(M) \to \operatorname{Tot}(N)$. If R_R is locally injective, then, in addition, for all $M \in$ Mod–R, $\operatorname{Tot}(M/\operatorname{Tot}(M)) = 0$.

3 Further properties of locally injective modules and locally projective modules

The definition of a locally injective, respectively locally projective, module was given by inner properties of the module. The characterization in Theorem 2.2 is by outer properties, i.e., by properties of Mod–R. In the following we give a further interesting characterization of locally injective modules by inner properties (Proposition 3.4.1).

In the following we use li for locally injective and lp for locally projective.

Proposition 3.1.

1) *Arbitrary direct sums of injective modules are* li.
2) *If* V *is* li *and* $V \subseteq^* M$*, then* M *is* li.
3) *Projective and semiperfect modules are* lp.

Proof. 1) Suppose that $V := \bigoplus_{i \in I} Q_i$ and each Q_i is injective. Assume that $A \subseteq V$, and A is not large in V. Then there exists $0 \neq C \subseteq V$ with $A \cap C = 0$. Choose $0 \neq C_0 \subseteq C$, C_0 finitely generated. Then also $A \cap C_0 = 0$. Since C_0 is finitely generated, there exists a finite set $I_0 \subseteq I$ such that $C_0 \subseteq Q := \bigoplus_{i \in I_0} Q_i$. Then Q is injective and Q contains an injective hull Q_0 of C_0, hence $0 \neq C_0 \subseteq^* \mathbb{Q}_0$. Since $A \cap C_0 = 0$, also $A \cap \mathbb{Q}_0 = 0$ (Lemma 0.5) and $\mathbb{Q}_0 \neq 0$. Hence V is li.

2) Assume $A \subseteq M$ and A not large in M. Then there exists $0 \neq C \subseteq M$ with $A \cap C = 0$. Since $V \subseteq^* M$ also $V \cap C \neq 0$ and $(A \cap V) \cap (V \cap C) \subseteq A \cap C = 0$. Hence $A \cap V$ is not large in V. By assumption there exists an injective $0 \neq Q \subseteq V$ with $(A \cap V) \cap Q = 0$. To show that $A \cap Q = 0$ we assume by way of contradiction that $A \cap Q \neq 0$. Since $V \subseteq^* M$, it follows that $V \cap (A \cap Q) \neq 0$ in contradiction to $(A \cap V) \cap Q = 0$. We conclude that $A \cap Q = 0$.

3) Suppose that P is projective, $B \subseteq P$, and B is not small in P. Then we consider $\nu : P \to P/B$. Since P is semiperfect by assumption, there exists a decomposition $P = P_1 \oplus P_2$, where the restriction ν_1 of ν to P_1 is a projective cover of P/B, i.e., $\mathrm{Ker}(\nu_1) \subseteq^\circ P_1$ and $P_2 \subseteq \mathrm{Ker}(\nu) = B$. If $P_2 = 0$, then $P_1 = P$ and $\mathrm{Ker}(\nu_1) = \mathrm{Ker}(\nu) = B \subseteq^\circ P$, a contradiction. Hence $P_2 \neq 0$ is the desired projective direct summand of P contained in B. □

Remark. *Locally injective modules need not be injective.* In fact, it is well–known (see e.g. [15, 6.5.1]) that R_R is Noetherian if and only if every direct sum of injective R–modules is injective. Hence if R_R is not Noetherian, then there exists an infinite direct sum of injective R–modules that is not injective, but is li by Proposition 3.1.1).

Corollary 3.2. *If* R_R *is Artinian, then* $\mathrm{Rad}(R) = \mathrm{Tot}(R)$.

Proof. By [15, 11.1.6, p. 278] we know that every projective Artinian module is semiperfect. Hence R_R is projective and semiperfect, and then by Proposition 3.1,

R_R is locally projective. By Corollary 2.5 it then follows that $\nabla(R_R, R_R) = \mathrm{Tot}(R_R, R_R) = \mathrm{Tot}(R)$. But

$$\nabla(R_R, R_R) = \{r \in R \mid rR \subseteq^\circ R_R\} = \mathrm{Rad}(R_R) = \mathrm{Rad}(R).$$

Together we have $\mathrm{Rad}(R) = \mathrm{Tot}(R)$. □

Proposition 3.3.

1) *Assume that V is* li *and satisfies the maximum condition for injective submodules. Then for every $A \subseteq V$, there exists an injective submodule $Q \subseteq V$ such that*

$$A \cap Q = 0, \quad A \oplus Q \subseteq^* V.$$

In particular, for $A = 0$ it follows that $Q = V$ is injective.

2) *Assume that W is* lp *and satisfies the maximum condition for projective direct summands of W. Then for every $B \subseteq W$, there exists a projective direct summand $P \subseteq^\oplus W$ and $U \subseteq^\circ W$ such that*

$$P \cap U = 0, \quad B = P \oplus U.$$

In particular, for $B = W$ it follows that $P = W$ is projective.

Proof. 1) Let Q be an injective submodule that is maximal with respect to $A \cap Q = 0$. Assume that $A \oplus Q$ is not large in V. Then there exists an injective submodule $Q_0 \subseteq V$ with $(A \oplus Q) \cap Q_0 = 0$, and then $Q + Q_0 = Q \oplus Q_0$ is again injective and $A \cap (Q \oplus Q_0) = 0$. This contradicts the maximality of Q, hence $A \oplus Q \subseteq^* V$. If $A = 0$, then $Q \subseteq^* V$; but, because of $G \subseteq^\oplus V$, this is possible only for $Q = V$.

2) Let P be a projective direct summand of W that is maximal with $P \subseteq B$. If $W = P \oplus C$, then $B = P \oplus (B \cap C)$. Assume that $B \cap C$ is not small in W. Then there exists a projective $P_0 \subseteq^\oplus W$, $P_0 \neq 0$, with $P_0 \subseteq B \cap C$. If $W = P_0 \oplus C_0$, then $B \cap C = P_0 \oplus (B \cap C \cap C_0)$. This implies that

$$B = P \oplus (B \cap C) = P \oplus P_0 \oplus (B \cap C \cap C_0).$$

To get a contradiction to the maximality of P, we still have to show that $P \oplus P_0$ is a direct summand of W. Since $P_0 \subseteq B \cap C \subseteq C$, it follows from $W = P_0 \oplus C_0$ that $C = P_0 \oplus (C \cap C_0)$ and further that $W = P \oplus C = P \oplus P_0 \oplus (C \cap C_0)$. If $W = P \oplus U$ with $U \subseteq^\circ W$, then $W = P$. □

In the following we call a set of submodules $\{U_i \mid i \in I\}$, $U_i \subseteq M$, **independent** if the sum of the U_i is direct.

Proposition 3.4.

1) *Assume that V is* li *and let $A \subseteq V$. Then there exists a maximal independent set $\{Q_i \mid i \in I\}$ of injective modules $Q_i \subseteq V$ with*

$$A \cap \left(\bigoplus_{i \in I} Q_i\right) = 0. \quad A \oplus \left(\bigoplus_{i \in I} Q_i\right) \subseteq^* V.$$

2) *Assume that W is* lp *and let $B \subseteq W$. Then there exists a maximal independent set $\{P_i \mid i \in I\}$ of projective modules $P_i \subseteq^{\oplus} W$, $P_i \subseteq B$ with the following property: If $C \subseteq B$ and*

$$\left(\bigoplus_{i\in I} P_i\right) \cap C = 0, \tag{1}$$

then $C \subseteq^{\circ} W$. This means that $P := \left(\bigoplus_{i\in I} P_i\right)$ is "nearly large" in B.

Proof. 1) Since the union of an ascending chain of independent sets $\{Q_i \mid i \in I\}$ of injective submodules $Q_i \subseteq V$ with

$$A \cap \left(\bigoplus_{i\in I} Q_i\right) = 0$$

is again such a set, we can apply Zorn's Lemma. Therefore we can assume that $\{Q_i \mid i \in I\}$ is maximal. But then $A \oplus \left(\bigoplus_{i\in I} Q_i\right)$ is large in V; otherwise there would exist an injective $Q \subseteq V$ with $\left(A \oplus \left(\bigoplus_{i\in I} Q_i\right)\right) \cap Q = 0$ contradicting the maximality of $\{Q_i \mid i \in I\}$.

2) Again by applying Zorn's Lemma we can assume that $\{P_i \mid i \in I\}$ is a maximal set of the desired kind. If for $C \subseteq B$ (1) is satisfied but C is not small in W, then there must exist a projective $0 \neq P \subseteq^{\oplus} W$ with $P \subseteq C$, contradicting the maximality of $\{P_i \mid i \in I\}$. □

Proposition 3.5.

1) *For a module V the following properties are equivalent.*

   i) *V is* li.

   ii) *There exists an independent set $\{Q_i \mid i \in I\}$ of injective modules $Q_i \subseteq V$ with*

$$\bigoplus_{i\in I} Q_i \subseteq^* V. \tag{2}$$

2) *If V is* li *and R_R is Noetherian, then V is injective and*

$$\Delta(V, M) = \mathrm{Rad}(V, M) = \mathrm{Tot}(V, M) \textit{ for all } M \in \text{ Mod-}R.$$

Proof. 1) i) $\Rightarrow$ ii): Proposition 3.4 with $A = 0$.

1) ii) $\Rightarrow$ i): Proposition 3.1.

2) If R_R is Noetherian, then the sum in (2) is injective, hence a direct summand of V. But a large direct summand must be the whole module. □

The following Proposition 3.6 constitutes a "weak form" of W being semiprime. For a semiprime module there is no restriction $x \notin \mathrm{Rad}(W)$. It is also a "weak form" of W being torsionless.

Proposition 3.6. *If W is* lp, *then for every $x \in W$ with $x \notin \mathrm{Rad}(W)$, there exists $f \in W^* := \mathrm{Hom}_R(W, R)$ such that*

$$xf(x) \neq 0. \tag{3}$$

Proof. Since $x \notin \text{Rad}(W)$, xR is not small in W. Hence there exists a projective module $P \neq 0$ that is a direct summand of W, $W = P \oplus C$, and $P \subseteq xR$. It follows that

$$xR = P \oplus (C \cap xR). \tag{4}$$

This implies that there are elements $s, t \in R$ such that

$$x = xs + xt, \quad xs \in P, \ xt \in C \cap xR. \tag{5}$$

Then $xsR \subseteq P$. If $p \in R$ such that $xp \in P$, then

$$xp = xsp + xtp.$$

Since (4) is a direct sum, this implies

$$xp = xsp, \ xtp = 0,$$

hence $P \subseteq xsR$. Together we have $xsR = P$. Now we consider the epimorphism

$$R \ni r \mapsto xsr \in P.$$

Since P is projective, this epimorphism splits. This means that there exists an idempotent $e \in R$, $e \neq 0$, such that

$$g : eR \ni er \mapsto xser \in P$$

is an isomorphism. By (5) it follows that

$$xe = xse + xte.$$

Since this sum is direct and $xse \neq 0$ (since $P \neq 0$), also $xe \neq 0$. Then we define $f \in W^*$ by

$$f \restriction_P = g^{-1}, \quad f \restriction_C = 0.$$

This implies

$$xf(xse) = xg^{-1}(xse) = xe \neq 0$$

and since

$$xf(xse) = xf(x)se$$

also $xf(x) \neq 0$. □

Finally in this section we show the following good properties of locally projective modules.

Theorem 3.7. *If M is locally projective, then*

1) $\text{Rad}(M) \subseteq^\circ M$;

2) $\text{Rad}(M) = \text{Tot}(M)$.

Proof. 1) Indirect. Assume that $\operatorname{Rad}(M)$ is not small in M. Then by definition of lp, there exists a non–zero projective direct summand P of M with $P \subseteq \operatorname{Rad}(M)$. Write $M = P \oplus A$. Then $\operatorname{Rad}(M) = P \oplus (A \cap \operatorname{Rad}(M))$ and also $\operatorname{Rad}(M) = \operatorname{Rad}(P) \oplus \operatorname{Rad}(A)$. Since $P \subseteq \operatorname{Rad}(M)$, $\operatorname{Rad}(P) \subseteq P$, and $\operatorname{Rad}(A) \subseteq A$, it follows by the Modular Law that

$$P = \operatorname{Rad}(P) \oplus (P \cap \operatorname{Rad}(A)) = \operatorname{Rad}(P).$$

But a non–zero projective module is not equal to its radical (see [15, p. 233]), a contradiction.

2) This is an easy consequence of the following special case of Theorem 2.2:

$$\nabla(R, M) = \operatorname{Tot}(R, M).$$

We apply to both sides of this equality the bijective mapping

$$\Phi : \operatorname{Hom}_R(R, M) \ni f \mapsto f(1) \in M.$$

First, if $f \in \nabla(R, M)$, then

$$\operatorname{Im}(f) = f(R) = f(1)R \subseteq^\circ M,$$

hence $f(1)R \subseteq \operatorname{Rad}(M)$. On the other hand, if $m \in \operatorname{Rad}(M)$, then the map f defined by

$$f(r) := mr, \quad r \in R$$

is in $\nabla(R, M)$. Therefore $\Phi(\nabla(R, M)) = \operatorname{Rad}(M)$. That furthermore $\Phi(\operatorname{Tot}(R, M)) = \operatorname{Tot}(M)$ follows directly by Definition II.1.2. □

It is an interesting fact that in the following sense the converse of Theorem 3.7 is satisfied for projective modules.

Theorem 3.8. *If P is a projective module with*

$$\operatorname{Rad}(P) \subseteq^\circ P \quad \textit{and} \quad \operatorname{Rad}(P) = \operatorname{Tot}(P),$$

then P is locally projective.

Proof. Let $A \subseteq P$ and suppose that A is not small in P. Then by assumption $A \not\subseteq \operatorname{Rad}(P) = \operatorname{Tot}(P)$. Hence A contains a pi–element a. There exists $\varphi \in \operatorname{Hom}_R(P, R)$ such that (in the sense of I.4., dual case):

$$d := a\varphi = d^2 \neq 0, \quad d \in \operatorname{End}(M).$$

Then $d(P) = a\varphi(P)$ is a non–zero direct summand of P, hence projective, and since $a \in A$ it is also contained in A. This shows that P is lp. □

We consider the following special case.

Corollary 3.9. *If R is a ring with $1 \in R$ and $\operatorname{Rad}(R) = \operatorname{Tot}(R)$, then R_R and ${}_RR$ are locally projective.*

Proof. R_R and ${}_RR$ are projective and since $R = 1R = R1$ is finitely generated, $\operatorname{Rad}(R)$ is small as a right– and left–ideal (Theorem 0.8). Hence all conditions of Theorem 3.8 are satisfied. □

By Theorem 1.2 and Theorem 2.2, for such a ring R with $\operatorname{Rad}(R) = \operatorname{Tot}(R)$ we have

$$\nabla(M, R) = \operatorname{Rad}(M, R) = \operatorname{Tot}(M, R)$$

for every $M \in$ Mod–R.

4 Totally good sets. Closure properties

As we have seen by Corollary II.1.10, the total is a semi–ideal in Mod–R, which means that it is closed under arbitrary multiplication from either side. But in general it is not closed under addition. For example, $\operatorname{Tot}(\mathbb{Z}) = \mathbb{Z} \setminus \{1, -1\}$. On the other hand, it is interesting that the radical $\operatorname{Rad}(\_,\_)$, the singular submodule Δ, and the co–singular submodule ∇ are contained in the total. These are ideals in Mod–R. If, for example, $\Delta(V, M) = \operatorname{Tot}(V, M)$, then also $\operatorname{Tot}(V, M)$ is additively closed. We have seen that there exist modules V and W such that

(i) $\Delta(V, M) = \operatorname{Tot}(V, M)$ for all $M \in$ Mod–R,

(ii) $\nabla(M, W) = \operatorname{Tot}(M, W)$ for all $M \in$ Mod–R,

(iii) $\operatorname{Rad}(V, M) = \operatorname{Tot}(V, M)$ for all $M \in$ Mod–R,

(iv) $\operatorname{Rad}(M, W) = \operatorname{Tot}(M, W)$ for all $M \in$ Mod–R.

However, we are not only interested in single modules V and W satisfying one of (i) through (iv), but we also intend to study the following "totally good sets".

$$\Omega = \Omega(R) := \{V \in \text{Mod–}R \mid V \text{ satisfies (i)}\},$$

$$\Psi = \Psi(R) := \{W \in \text{Mod–}R \mid W \text{ satisfies (ii)}\},$$

$$\Phi = \Phi(R) := \{V \in \text{Mod–}R \mid V \text{ satisfies (iii)}\},$$

$$\Gamma = \Gamma(R) := \{W \in \text{Mod–}R \mid W \text{ satisfies (iv)}\},$$

$$\mathfrak{J} = \mathfrak{J}(R) := \{V \in \text{Mod–}R \mid V = \bigoplus \text{injectives}\}$$

$$\mathfrak{I} = \mathfrak{I}(R) := \{Q \in \text{Mod–}R \mid Q = \text{ injective}\},$$

$$\mathfrak{P} = \mathfrak{P}(R) := \{P \in \text{Mod–}R \mid P = \text{ projective \& semiperfect}\}.$$

First we show that Ω, Ψ, Φ, Γ are closed under taking direct summands and finite direct sums. For the proofs we have to use that Tot is a semi–ideal and that Rad, Δ, and ∇ are ideals in Mod–R. For the sake of completeness we prove this for Δ and ∇.

Lemma 4.1. *Δ and ∇ are ideals in* Mod–R.

Proof. Additive closure property. Assume that $f, g \in \Delta(M, W)$. Then

$$\operatorname{Ker}(f) \cap \operatorname{Ker}(g) \subseteq \operatorname{Ker}(f+g).$$

Since the intersection of two large submodules is large, also the larger submodule $\operatorname{Ker}(f+g)$ is large in M. Thus $f+g \in \Delta(M, W)$.

Assume next that $f, g \in \nabla(M, W)$. Then

$$\operatorname{Im}(f+g) \subseteq \operatorname{Im}(f) + \operatorname{Im}(g)$$

and since the sum of two small submodules is small, also the smaller submodule $\operatorname{Im}(f+g)$ is small in W. Thus $f+g \in \Delta(M, W)$.

Multiplicative closure property. For any $A \in$ Mod–R, $f \in \Delta(M, W)$, and $h \in \operatorname{Hom}_R(W, A)$,

$$\operatorname{Ker}(f) \subseteq \operatorname{Ker}(hf).$$

Since $\operatorname{Ker}(f) \subseteq^* W$, also $\operatorname{Ker}(hf) \subseteq^* W$, hence $hf \in \Delta(M, A)$. Let next $f \in \Delta(M, W)$ and $h \in \operatorname{Hom}_R(A, M)$. We must show that $fh \in \Delta(A, W)$. Consider $0 \neq a \in A$. If $a \in \operatorname{Ker}(h)$, then $a \in \operatorname{Ker}(fh)$. If $a \notin \operatorname{Ker}(h)$, then $h(a) \neq 0$, and then the assumption that $\operatorname{Ker}(f)$ is large implies that there exists $r \in R$ such that $0 \neq h(a)r \in \operatorname{Ker}(f)$. But then $fh(ar) = 0$, hence $ar \in \operatorname{Ker}(fh)$. This implies $\operatorname{Ker}(fh) \subseteq^* A$ and $fh \in \Delta(A, W)$.

We turn to the multiplicative closure of ∇. Let $f \in \nabla(M, W)$ and $h \in \operatorname{Hom}_R(A, M)$. Then

$$\operatorname{Im}(fh) \subseteq \operatorname{Im}(f) \subseteq^\circ W,$$

hence $\operatorname{Im}(fh) \subseteq^\circ W$ and $fh \in \nabla(A, W)$.

Let $f \in \nabla(M, W)$ and $h \in \operatorname{Hom}_R(W, A)$. Then assume that for $U \subseteq A$,

$$\operatorname{Im}(hf) + U = A.$$

It then follows ($h^{-1}(C)$ is the pre–image of $C \subseteq A$) that

$$h^{-1}(\operatorname{Im}(fg)) + h^{-1}(U) = h^{-1}(A) = W = \operatorname{Im}(f) + h^{-1}(U).$$

Since $\operatorname{Im}(f) \subseteq^\circ W$, it follows that $h^{-1}(U) = W$, hence $\operatorname{Im}(h) \subseteq U$, and this implies $\operatorname{Im}(hf) \subseteq U$. Therefore $hf \in \nabla(M, A)$. □

In the proof of the following theorem we use the semi–ideal property of Tot and the ideal properties of Δ, ∇ and Rad without explicit mention.

Furthermore, we use the following notations: $\pi : A \oplus B \to A$ denotes the projection of $A \oplus B$ onto A along B, and ι denotes the inclusion of A in $A \oplus B$. If we have direct sums $V_1 \oplus V_2$ (or $W_1 \oplus W_2$), then π_i denotes the projection of the direct sum onto the ith summand along the other summand, and ι_i denotes the inclusion of the ith summand in the direct sum. Then $1_i = 1_{V_i} = \pi_i \iota_i$, $e_i = \iota_i \pi_i$ is an idempotent and $e_1 + e_2 = 1_{V_1 \oplus V_2}$.

Theorem 4.2. *The families Ω, Ψ, Φ, Γ are closed under taking direct summands and finite direct sums.*

Proof for Ω. Let $V \in \Omega$ and $V = A \oplus B$. Since always $\Delta(A, M) \subseteq \mathrm{Tot}(A, M)$, we only have to show $\mathrm{Tot}(A, M) \subseteq \Delta(A, M)$. Assume $f \in \mathrm{Tot}(A, M)$. Then $f\pi \in \mathrm{Tot}(V, M) = \Delta(V, M)$. This implies

$$f\pi\iota = f1_A = f \in \Delta(A, M),$$

which was to be shown.

Now consider $V_1, V_2 \in \Omega$ and $f \in \mathrm{Tot}(V_1 \oplus V_2, M)$. Then it follows that

$$f\iota_1 \in \mathrm{Tot}(V_1, M) = \Delta(V_1, M), \quad f\iota_2 \in \mathrm{Tot}(V_2, M) = \Delta(V_2, M).$$

Since Δ is an ideal, these imply

$$f\iota_1\pi_1 = fe_1, \quad f\iota_2\pi_2 = fe_2 \in \Delta(V_1 \oplus V_2, M)$$

and

$$fe_1 + fe_2 = f(e_1 + e_2) = f \in \Delta(V_1 \oplus V_2, M).$$

Proof for Ψ. Assume $W \in \Psi$, $W = A \oplus B$ and $f \in \mathrm{Tot}(M, A)$. Then $\iota f \in \mathrm{Tot}(M, W) = \nabla(M, W)$. Then it follows that

$$\pi\iota f = 1_A f = f \in \nabla(M, A),$$

which was to be proved.

Now we consider $W_1, W_2 \in \Psi$ and let $f \in \mathrm{Tot}(M, W_1 \oplus W_2)$. Then $\pi_i f \in \mathrm{Tot}(M, W_i) = \nabla(M, W_i)$, $i = 1, 2$. These imply

$$\iota_i\pi_i f = e_i f \in \nabla(M, W_1 \oplus W_2)$$

and

$$e_1 f + e_2 f = (e_1 + e_2)f = f \in \nabla(M, W_1 \oplus W_2).$$

Proof for Φ. Let $V \in \Phi$, $V = A \oplus B$, and $f \in \mathrm{Tot}(A, M)$. Then

$$f\pi \in \mathrm{Tot}(V, M) = \mathrm{Rad}(V, M)$$

and

$$f\pi\iota = f \in \mathrm{Rad}(A, M).$$

Now assume that $V_1, V_2 \in \Phi$ and $f \in \mathrm{Tot}(V_1 \oplus V_2, M)$. Then

$$f\iota_i \in \mathrm{Tot}(V_i, M) = \mathrm{Rad}(V_i, M)$$

and this implies that

$$f\iota_i\pi_i = fe_i \in \mathrm{Rad}(V_1 \oplus V_2, M)$$

and $fe_1 + fe_2 = f(e_1 + e_2) = f \in \mathrm{Rad}(V_1 \oplus V_2, M)$.

Proof for Γ. Similar to the preceding proofs. □

What do we know about the sets Ω, Ψ, Φ, Γ besides their closure properties?

The sets Ω and Ψ. Here Theorem 2.2 provides satisfactory information. The elements of Ω are exactly the locally injective modules and the elements of Ψ are exactly the locally projective modules. By Proposition 3.1 and Proposition 3.5 we also know that

$$\mathfrak{I} \subseteq \mathfrak{J} \subseteq \Omega, \quad \mathfrak{P} \subseteq \Psi.$$

The sets Φ and Γ. Here we do not have a similar "inner" characterization as for Ω and Ψ. By Corollary 2.5 we know: If V is large restricted and locally injective, then $V \in \Phi$. Dually: If W is small restricted and locally projective, then $W \in \Gamma$. These imply:

$$\mathfrak{J} \subseteq \Phi, \quad \mathfrak{P} \subseteq \Gamma.$$

These results motivate the following questions. Are the conditions LR and li together also necessary for the modules in Φ? Are the conditions SR and lp together also necessary for the modules in Γ? Furthermore, one can ask: Is a direct sum of injective modules LR?

5 Relatively regular elements

Definition 5.1. Let $\emptyset \neq U \subseteq \mathrm{Hom}_R(M, W)$. Then $f \in \mathrm{Hom}_R(M, W)$ is called U–**regular** if and only if there exists $g \in \mathrm{Hom}_R(M, W)$ such that

$$fgf - f \in U.$$

If f is U–regular and $U = \{0\}$, then f is regular.

We are interested in finding $\mathrm{Tot}(M, W)$–regular elements. In order to do this we need "strong" assumptions. The obvious question arises whether weaker assumptions are sufficient.

For the proofs we have to use the following well–known properties of injective and of projective and semiperfect modules.

(C1) Let $A \subseteq V$. If B is a complement of A in V and A_1 a complement of B in V with $A \subseteq A_1$, then

$$A + B \subseteq^* V, \quad V = A_1 \oplus B.$$

(D1) Let $A \subseteq W$. If B is a supplement of A in W and A_0 a supplement of B in W with $A_0 \subseteq A$, then

$$A \cap B \subseteq^\circ W, \quad W = A_0 \oplus B.$$

Remark. If V is injective, then V satisfies (C1). If W is projective and semiperfect, then W satisfies (D1). Proofs of these facts can be found in [26] where the notations (C1) and (D1) are introduced but proofs can also be found in other books, e.g. [15].

Theorem 5.2.

1) *If Q is injective, then for every $M \in \text{Mod-}R$ every map $f \in \text{Hom}_R(Q, M)$ is $\Delta(Q, M)$–regular.*

2) *If P is projective and semiperfect, then for every $M \in \text{Mod-}R$, every map $f \in \text{Hom}_R(M, P)$ is $\nabla(M, P)$–regular.*

Remark. If Q is injective, then by Corollary 2.5,

$$\Delta(Q, M) = \text{Rad}(Q, M) = \text{Tot}(Q, M).$$

If P is projective and semiperfect, then by Corollary 2.5,

$$\nabla(M, P) = \text{Rad}(M, P) = \text{Tot}(M, P).$$

Proof of Theorem 5.2. 1) If $f \in \Delta(Q, M)$, then for every $g \in \text{Hom}_R(M, Q)$,

$$fgf - f = (fg - 1_M)f \in \Delta(Q, M).$$

Assume now that $f \in \text{Hom}_R(Q, M)$ and $\text{Ker}(f)$ is not large in Q. Apply (C1) with $A = \text{Ker}(f)$ to get

$$\text{Ker}(f) \subseteq A_1, \quad \text{Ker}(f) + B \subseteq^* Q, \quad Q = A_1 \oplus B. \tag{6}$$

Since $\text{Ker}(f)$ is not large in Q, the complement B of $\text{Ker}(f)$ is non–zero. As $\text{Ker}(f) \cap B = 0$, the mapping

$$\varphi : B \ni b \mapsto f(b) \in f(B)$$

is an isomorphism. Since B is injective, also $f(B)$ is injective and a direct summand of M. Denote by π some projection of M onto $f(B)$ and by $\iota : B \to Q$ the inclusion. Then, with $g := \iota\varphi^{-1}\pi \in \text{Hom}_R(M, Q)$, it follows that for every $b \in B$,

$$fgf(b) = fg(f(b)) = f(b).$$

This implies

$$\text{Ker}(f) + B \subseteq \text{Ker}(fgf - f),$$

and hence, by (6), $fgf - f \in \Delta(Q, M)$.

2) Similarly, if $f \in \nabla(M, P)$, and $g \in \text{Hom}_R(P, M)$, then

$$fgf - f = f(gf - 1_M) \in \nabla(M, P).$$

Assume now that $f \in \text{Hom}_R(M, P)$ and $\text{Im}(f)$ is not small in P. In (D1) with $A = \text{Im}(f)$, the supplement B of $\text{Im}(f)$ is not P, and so $A_0 \neq 0$. Denote by $\pi : P \to A_0$ the projection belonging to $P = A_0 \oplus B$. Then $\text{Ker}(\pi) = B$. Since $A_0 \subseteq \text{Im}(f)$, the mapping

$$\pi f : M \to A_0$$

is an epimorphism onto the projective module A_0, hence it is splitting:

$$M = N \oplus \operatorname{Ker}(\pi f).$$

Consequently,

$$\varphi : N \ni x \mapsto \pi f(x) \in A_0$$

is an isomorphism. If $\iota : N \to M$ is the inclusion, we define $g := \iota\varphi^{-1}\pi$. For $x \in N$ it follows that

$$fgf(x) = f\iota\varphi^{-1}\pi f(x) = f(x),$$

hence $(fgf - f)(N) = 0$. For $y \in \operatorname{Ker}(\pi f)$,

$$fgf(y) = f\iota\varphi^{-1}(\pi f(y)) = 0,$$

thus, using (D1),

$$(fgf - f)(\operatorname{Ker}(\pi f)) = -f(\operatorname{Ker}(\pi f)) = \operatorname{Im}(f) \cap \operatorname{Ker}(\pi) = \operatorname{Im}(f) \cap B \subseteq^{\circ} P.$$

Together we have $fgf - f \in \nabla(M, P)$. □

In the special case $Q = M$ and $M = P$, the full assumption that Q is injective and that P is projective and semiperfect, respectively, is not needed. The following concepts appear in [26].

A module V is called **continuous** if, in addition to (C1) it satisfies

(C2) If a submodule $A \subseteq V$ is isomorphic to a direct summand of V, then A is a direct summand of V.

A module W is called **discrete** if in addition to (D1) it satisfies

(D2) If A is a submodule of W such that W/A is isomorphic to a direct summand of W, then A is a direct summand of W.

Corollary 5.3 (Special case).

1) *If V is continuous, then every $f \in \operatorname{Hom}_R(V, V)$ is $\Delta(V, V)$–regular.*

2) *If W is discrete, then every $f \in \operatorname{Hom}_R(W, W)$ is $\nabla(W, W)$–regular.*

Proof. 1) In the proof of Theorem 5.2 we used the isomorphism φ. It now follows by (C2) that $f(B)$ is a direct summand of V. The remaining parts to the proof are the same.

2) In the proof of Theorem 5.2 the map πf was an epimorphism that splits. Now, $W/\operatorname{Ker}(\pi f) \cong A_0$, which is a direct summand. By (D2) it follows that $\operatorname{Ker}(\pi f)$ is a direct summand, hence $W = N \oplus \operatorname{Ker}(\pi f)$. The remaining steps of the proof are unchanged. □

6 Modules with “good” endomorphism rings and exchange properties

We now consider rings that have “good” properties with respect to the total. This means that the total is equal to the radical or at least additively closed, i.e., a two–sided ideal. Mainly we consider modules that have “good” endomorphism rings in this sense. Interesting examples of such modules are the exchange modules considered in the second part of this section.

For rings that have a “good” total we have special notations.

Definition 6.1.

1) A ring S is called a **total ring** if and only if $\mathrm{Tot}(S)$ is additively closed, and hence a two–sided ideal in S.

2) S is a **radical total ring** if and only if $\mathrm{Rad}(S) = \mathrm{Tot}(S)$.

3) A module M is called a **total module** or **radical total module** if and only if $S := \mathrm{End}(M)$ is a total ring or a radical total ring, respectively.

We get examples for these notions by specializing our general results.

Example 6.2. Let $S = \mathrm{End}(M)$.

1) If M is locally injective, then $\Delta(M, M) = \mathrm{Tot}(M, M) = \mathrm{Tot}(S)$, hence M is a total module.

2) If M is locally projective, then $\nabla(M, M) = \mathrm{Tot}(M, M) = \mathrm{Tot}(S)$, hence M is a total module.

3) If M is locally injective and LR (= large restricted), then M is a radical total module.

4) If M is locally projective and SR (= small restricted), then M is a radical total module.

Proof. 1) and 2) are special cases of Theorem 2.2 while 3) and 4) are special cases of Corollary 2.5. □

Later we will see that exchange modules are further interesting examples.

Now, we show that if M is a total or radical total module, the same is true for direct summands of M. The first part of the following lemma is well–known, but we could not find a reference for the proof. Therefore we give the proof in detail.

Lemma 6.3. *Let S be a ring with $1 \in S$, and let e be an idempotent in S. Then*

1) $e\,\mathrm{Rad}(S)e = \mathrm{Rad}(S) \cap eSe = \mathrm{Rad}(eSe)$.

2) $e\,\mathrm{Tot}(S)e = \mathrm{Tot}(S) \cap eSe = \mathrm{Tot}(eSe)$.

Proof. We use the fact that eSe is a ring with identity $e = e1e$.

$e\,\mathrm{Rad}(S)e \subseteq \mathrm{Rad}(S) \cap eSe$: This follows from the fact that $\mathrm{Rad}(S)$ is a two–sided ideal in S.

$\mathrm{Rad}(S) \cap eSe \subseteq \mathrm{Rad}(eSe)$: Let $u \in \mathrm{Rad}(S) \cap eSe$. Then $eu = ue = u$. We show that $ueSe \subseteq^\circ eSe$. Assume that

$$eSe = B + ueSe,$$

where $B \subseteq eSe$ is a right ideal in eSe. It follows that

$$e = b + uese, \quad b \in B \text{ with } eb = b, s \in S.$$

With this we get

$$S = eS + (1-e)S = bS + ueseS + (1-e)S.$$

Since $u \in \mathrm{Rad}(S)$, we conclude that $ueseS \subseteq^\circ S_S$, hence $S = bS \oplus (1-e)S$. Multiplying on the left by e we get

$$eS = ebS = bS,$$

hence $e \in B$ and $B = eSe$. This means that $ueSe \subseteq^\circ eSe$, thus $u \in \mathrm{Rad}(eSe)$.

$\mathrm{Rad}(eSe) \subseteq e\,\mathrm{Rad}(S)e$: Let $u \in \mathrm{Rad}(eSe)$. Then again $eu = ue = u$. Assume

$$S = B + uS, \quad B \subseteq S_S.$$

Then it follows that

$$e = b_0 + us_0, \; b_0 \in B, s_0 \in S, \tag{7}$$

$$1 = b_1 + us_1, \; b_1 \in B, s_1 \in S. \tag{8}$$

Since $eu = u$ (7) implies that $eb_0 = b_0$ and

$$eSe = b_0eSe + us_0eSe.$$

Since $u \in \mathrm{Rad}(eSe)$ and $b_0eSe = eb_0eSe \subseteq eSe_{eSe}$, we get

$$eSe = b_0eSe,$$

hence $e = b_0ese \in B$ where $s \in S$. Since $(1-e)u = 0$, it follows from (8) that

$$1 - e = (1-e)b_1 = b_1 - eb_1,$$

and since $b_1 \in B$ and $e \in B$, also $1 - e \in B$. Together we have $1 \in B$ which means $B = S$. Therefore $uS \subseteq^\circ S_S$, hence $u \in \mathrm{Rad}(S)$.

This proves 1).

2) We prove first that

$$ese \text{ is pi in } eSe \Leftrightarrow ese \text{ is pi in } S. \tag{9}$$

$\Rightarrow$: Clear since an idempotent in eSe is also an idempotent in S.

$\Leftarrow$: Assume that

$$d := eset = d^2 \neq 0, \ t \in S.$$

Then $ed = d$. This implies

$$(ede)^2 = edeede = edede = ed^2e = ede$$

and

$$d(ede)d = d^3 = d \neq 0.$$

Hence

$$ede = (ese)(ete),$$

showing that ese is pi in eSe.

Now, by (9), it follows that

$$ese \in \mathrm{Tot}(eSe) \Leftrightarrow ese \in \mathrm{Tot}(S). \tag{10}$$

To prove 2) we show first that $e\,\mathrm{Tot}(S)e \subseteq \mathrm{Tot}(S) \cap eSe$. Since $\mathrm{Tot}(S)$ is closed under multiplication from both sides, $e\,\mathrm{Tot}(S)e \subseteq \mathrm{Tot}(S)$ and also $e\,\mathrm{Tot}(S)e \subseteq eSe$.

$\mathrm{Tot}(S) \cap eSe \subseteq \mathrm{Tot}(eSe)$ follows by (10).

$\mathrm{Tot}(eSe) \subseteq e\,\mathrm{Tot}(S)e$. If $ese \in \mathrm{Tot}(eSe)$, then, by (10), $ese \in \mathrm{Tot}(S)$ and therefore $ese = e(ese)e \in e\,\mathrm{Tot}(S)e$. □

Corollary 6.4. *Let S be a ring with $1 \in S$ and let $e \in S$ be an idempotent. Then the following statements hold.*

1) *If S is a total ring, then eSe is a total ring.*
2) *If S is a radical total ring, then eSe is a radical total ring.*
3) *If $\mathrm{Tot}(S) = 0$, then $\mathrm{Tot}(eSe) = 0$.*

Proof. 1) If $es_1e, es_2e \in \mathrm{Tot}(eSe)$, then, by Lemma 6.3.2), both are in $\mathrm{Tot}(S)$. By assumption we have further that $es_1e + es_2e \in \mathrm{Tot}(S)$. Since

$$es_1e + es_2e = e(es_1e + es_2e)e,$$

again by Lemma 6.3.2), the element is in $\mathrm{Tot}(eSe)$.

2) If $\mathrm{Rad}(S) = \mathrm{Tot}(S)$, then by Lemma 6.3.1) and 6.3.2) it follows that

$$e\,\mathrm{Rad}(S)e = \mathrm{Rad}(eSe) = e\,\mathrm{Tot}(S)e = \mathrm{Tot}(eSe).$$

3) Lemma 6.3.2). □

We intend to show next that total and radical total modules transmit these properties to direct summands. These facts are also used in connection with the exchange properties.

Consider

$$A = B \oplus C, \tag{11}$$

and denote by

$$\iota : B \to A \text{ the inclusion}, \qquad \pi : A \to B \text{ the projection}$$

belonging to (11). Then

$$\begin{array}{lll} \pi\iota = 1_B & \text{and} & \iota\pi\iota = \iota 1_B = \iota, \\ \iota\pi =: e = e^2 & \text{and} & \pi\iota\pi = \pi e = 1_B\pi = \pi. \end{array}$$

Thus e is the projector onto B belonging to (11). Setting $S := \mathrm{End}(A)$, we will see that

$$\pi S\iota = \mathrm{End}(B).$$

First, it is clear that $\pi S\iota \subseteq \mathrm{End}(B)$. Suppose that $\varphi \in \mathrm{End}(B)$, Then $\iota\varphi\pi \in S$ and $\pi(\iota\varphi\pi)\iota = 1_B\varphi 1_B = \varphi$, hence also $\mathrm{End}(B) \subseteq \pi S\iota$.

If one is interested in the connection between $S = \mathrm{End}(A)$ and $\mathrm{End}(B)$, then it is technically easier to work not with $\mathrm{End}(B)$ itself but with the isomorphic subring eSe of S. The relevant ring isomorphism is explicitly

$$\mathrm{End}(B) = \pi S\iota \ni \pi f\iota \mapsto \iota\pi f\iota\pi = efe \in eSe \tag{12}$$

with inverse isomorphism

$$eSe \ni efe \mapsto \pi efe\iota = \pi fe \in \mathrm{End}(B). \tag{13}$$

From Corollary 6.4 it obviously follows:

Corollary 6.5. *Direct summands of total and radical total modules have the same property respectively.*

In the following we will show that exchange properties for modules are good conditions for the total of their endomorphism ring.

We start with the definitions of two exchange properties, the first of which is the classical 2–exchange property. Again let R be a ring with $1 \in R$ and denote by $A, B, C, \ldots$ unitary right R–modules.

Definition 6.6.

1) The module A has the 2–**exchange property**, (= 2–EP) if and only if whenever

$$M = A \oplus B = C \oplus D, \tag{14}$$

then there exist $C' \subseteq C$, $D' \subseteq D$ such that

$$M = A \oplus C' \oplus D'. \tag{15}$$

2) The module A has the **D2–exchange property** (= D2–EP) if and only if whenever $0 \neq A_0 \subseteq^{\oplus} A$ and

$$M = A_0 \oplus B = C \oplus D$$

there exist $0 \neq A_0' \subseteq A_0$, $C' \subseteq C$, $D' \subseteq D$ such that

$$M = A_0' \oplus C' \oplus D'.$$

It is obvious that these definitions are preserved under module isomorphisms. The "D" in the second definition is a hint to direct summands. Examples of modules with exchange properties are given at the end of this chapter.

Before we start with details we mention the main result.

Main Theorem 6.7.

1) *If A has the* 2–EP, *then A is a radical total module.*

2) *If A has the* D2–EP, *then A is a total module.*

The formulation of the main theorem is short and easy to understand but the proof is long and somewhat tricky. We start with two lemmas.

Lemma 6.8.

1) *Any direct summand of a module with the* 2–EP *also has the* 2–EP.

2) *The* 2–EP *implies* D2–EP.

Proof. 1) Assume that A has the 2–EP and

$$A = A_1 \oplus A_2.$$

In order to show that A_1 also has the 2–EP, we assume

$$M = A_1 \oplus B = C \oplus D.$$

Then we consider the external direct sum

$$M \overset{\bullet}{\oplus} A_2 = \{(m, a_2) \mid m \in M, a_2 \in A_2\}.$$

Setting $A_3 = (0, A_2)$ and identifying A_1, B, C, D with $(A_1, 0)$, $(B, 0)$, $(C, 0)$, $(D, 0)$, respectively, we obtain the internal decomposition

$$M \overset{\bullet}{\oplus} A_2 = A_1 \oplus B \oplus A_3 = C \oplus D \oplus A_3. \tag{16}$$

Also

$$A_2 \cong A_3, \quad A_1 \oplus A_3 \cong A_1 \oplus A_2 = A.$$

Now the 2–EP of $A_1 \oplus A_3 \cong A$ will be applied to the decomposition (16) in the form

$$M \overset{\bullet}{\oplus} A_2 = (A_1 \oplus A_3) \oplus B = (A_3 \oplus C) \oplus D.$$

Accordingly, there exist $U \subseteq A_3 \oplus C$, $D' \subseteq D$ such that

$$M \overset{\bullet}{\oplus} A_2 = (A_1 \oplus A_3) \oplus U \oplus D'. \tag{17}$$

From $A_3 \subseteq A_3 \oplus U \subseteq A_3 \oplus C$ it follows by the Modular Law that

$$A_3 \oplus U = (A_3 \oplus U) \cap (A_3 \oplus C) = A_3 \oplus (C \cap (A_3 \oplus U)). \tag{18}$$

Set

$$C' := C \cap (A_3 \oplus U) \subseteq C.$$

Then (17) and (18) together imply

$$M \overset{\bullet}{\oplus} A_2 = A_1 \oplus A_3 \oplus C' \oplus D'. \tag{19}$$

Recalling that $A_3 = (0, A_2)$, the projection π of $M \oplus A_2$ onto M along A_2 then produces by (16) and (19) the desired decomposition

$$M = A_1 \oplus B = A_1 \oplus C' \oplus D'.$$

2) If A has the 2–EP and if $0 \neq A_0 \subseteq^{\oplus} A$, then A_0 has the 2–EP. In fact, let

$$M = A_0 \oplus B = C \oplus D,$$

then there exist $C' \subseteq C$, $D' \subseteq D$ with

$$M = A_0 \oplus C' \oplus D',$$

hence the D2–EP is satisfied (even with $A_0' = A_0$). □

A non–zero direct summand of a module with the D2–EP has again the D2–EP. This follows directly from the definition.

We now come to the lemma that provides for a certain flexibility with direct summands.

Lemma 6.9. *Let*

$$M = C \oplus D = H \oplus C' \oplus D', \quad C' \subseteq C, \ D' \subseteq D. \tag{20}$$

Then there exists a decomposition $H = H' \oplus H''$ *such that*

$$M = H' \oplus C \oplus D' = H'' \oplus C' \oplus D \tag{21}$$

and if $D' \neq D$, *then* $H' \neq 0$, *if* $C' \neq C$, *then* $H'' \neq 0$.

Proof. By the Modular Law applied to (20) it follows that C' and D' are direct summands of C and D respectively. Write

$$C = C' \oplus C'', \quad D = D' \oplus D''$$

and obtain

$$M = C' \oplus C'' \oplus D' \oplus D'' = H \oplus C' \oplus D'. \tag{22}$$

By Lemma 0.4 there exists $\varphi \in \mathrm{Hom}(C'' \oplus D'', C' \oplus D')$ such that

$$H = (\varphi + 1)(C'' \oplus D'') = (\varphi + 1)(C'') \oplus (\varphi + 1)(D'').$$

Set $H' := (\varphi + 1)(C'') \subseteq H$ and $H'' := (\varphi + 1)(D'')$ and observe that, again by Lemma 0.4 applied to the restrictions of φ to C'' and D'' respectively, it follows that

$$C'' \oplus C' \oplus D' = (\varphi + 1)(C'') \oplus C' \oplus D' = H' \oplus C' \oplus D',$$
$$D'' \oplus C' \oplus D' = (\varphi + 1)(D'') \oplus C' \oplus D' = H'' \oplus C' \oplus D'.$$

Substituting in (22) it follows that

$$M = H' \oplus C' \oplus D = H'' \oplus C \oplus D'.$$

The remaining claims are immediate. $\square$

In order to establish the connection between endomorphisms of A and exchange properties of A, we require the following special modules.

Let A_R be a given module and let $f \in \mathrm{End}(A_R)$. Further let

$$\begin{aligned} M &:= A \dot{\oplus} A = \{(a_1, a_2) \mid a_1, a_2 \in A\}, \\ A_1 &:= \{(a, 0) \mid a \in A\}, \\ A_2 &:= \{(0, a) \mid a \in A\}, \\ C &:= \{(f(a), f(a) - a) \mid a \in A\}, \\ D &:= \{(a, a) \mid a \in A\}. \end{aligned}$$

In addition we need the following homomorphisms.

$$\begin{aligned} \alpha_1 &: A \ni a \mapsto (a, 0) \in A_1, \\ \alpha_2 &: A \ni a \mapsto (0, a) \in A_2, \\ \gamma &: A \ni a \mapsto (f(a), f(a) - a) \in C, \\ \delta &: A \ni a \mapsto (a, a) \in D. \end{aligned}$$

It is obvious that α_1, α_2 and δ are isomorphisms, but also γ is an isomorphism. To verify this we only have to check injectivity. Assume that $(f(a), f(a) - a) = (0, 0)$. Then $f(a) = 0$ and $a = f(a) = 0$.

We have $M = A_1 \oplus A_2$, but also $M = C \oplus D$ is true: For $a_1, a_2 \in A$,

$$\begin{aligned}(a_1, a_2) &= (f(a_1 - a_2), f(a_1 - a_2) - (a_1 - a_2)) \\ &\quad + (a_1 - f(a_1 - a_2), a_1 - f(a_1 - a_2)),\end{aligned}$$

hence $M = C + D$. Assume that

$$(f(a), f(a) - a) = (a', a') \in C \cap D.$$

Then $f(a) = a'$, $f(a) - a = a'$, and this implies that $a = 0$, $a' = 0$. Altogether we have

$$M = A_1 \oplus A_2 = C \oplus D, \quad A \cong A_1 \cong A_2 \cong C \cong D.$$

We will use these properties without explicit reference.

In the following lemma the assumptions are consequences of the exchange properties. Keeping this in mind the meaning of the assumptions is better understood.

Lemma 6.10. *Let A be an R–module and $f \in \mathrm{End}(A)$. The modules M, A_1, A_2, C, D are the modules introduced above. In particular $M = A_1 \oplus A_2 = C \oplus D$, and f enters via $C = \{(f(a), f(a) - a) \mid a \in A\}$. Then the following statements hold.*

1) *If $M = C \oplus A_1' \oplus A_2'$ with $A_1' \subseteq A_1$, $A_2' \subseteq A_2$, and $A_2' \neq 0$, then f is* pi *(in* $\mathrm{End}(A)$*).*

2) *If $M = C' \oplus A_1' \oplus A_2$ with $A_1' \subseteq A_1$, $C' \subseteq C$ and $C' \neq 0$, then f is* pi.

3) *If $M = C' \oplus A_1 \oplus A_2'$ with $A_2' \subseteq A_2$, $C' \subseteq C$ and $C' \neq 0$, then $1_A - f$ is* pi.

Proof. 1) For the proof we use Lemma II.1.1.4), that is, we will show that f induces an isomorphism between non–zero direct summands of A.

By assumption and since $A_2 \subseteq M$, it follows with the Modular Law that

$$A_2 = A_2' \oplus (A_2 \cap (C \oplus A_1')).$$

Set $A_2'' := A_2 \cap (C \oplus A_1')$. Then

$$A = \alpha_2^{-1}(A_2) = \alpha_2^{-1}(A_2') \oplus \alpha_2^{-1}(A_2'') \tag{23}$$

and since $A_2' \neq 0$, also $\alpha_2^{-1}(A_2') \neq 0$. Let π_D be the projection of $M = C \oplus D$ onto D along C. Since $\mathrm{Ker}(\pi_D) = C$, and $M = C \oplus A_1' \oplus A_2'$, the map π_D induces an isomorphism of $A_1' \oplus A_2'$ with D. Therefore

$$D = \pi_D(A_1') \oplus \pi_D(A_2'), \quad \pi_D(A_2') \neq 0.$$

Thus

$$A = \delta^{-1}(D) = \delta^{-1}\pi_D(A_1') \oplus \delta^{-1}\pi_D(A_2'), \quad \delta^{-1}\pi_D(A_2') \neq 0. \tag{24}$$

We will prove that f induces an isomorphism of $\alpha_2^{-1}(A_2')$ with $\delta^{-1}\pi_D(A_2')$ via

$$\alpha_2^{-1}(A_2') \xrightarrow{\widehat{\alpha_2}} A_2' \xrightarrow{\widehat{\pi_D}} \pi_D(A_2') \xrightarrow{\widehat{\delta^{-1}}} \delta^{-1}\pi_D(A_2')$$

with isomorphisms $\widehat{\alpha_2}$, $\widehat{\pi_D}$, $\widehat{\delta^{-1}}$ induced by α_2, π_D, δ^{-1}. For $x \in \alpha_2^{-1}(A_2')$ we have

$$\alpha_2(x) = (0, x) = (f(-x), f(-x) + x) + (f(x), f(x)) \tag{25}$$

with $(f(-x), f(-x) + x) \in C$, $(f(x), f(x)) \in D$. It follows that

$$\pi_D(\alpha_2(x)) = \pi_D(0, x) = (f(x), f(x))$$

and

$$\delta^{-1}\pi_D(0, x) = \delta^{-1}(f(x), f(x)) = f(x),$$

hence, in combination,

$$\alpha_2^{-1}(A_2') \ni x \mapsto f(x) \in \delta^{-1}\pi_D(A_2') \tag{26}$$

is an isomorphism. Since $A_2' \neq 0$, also $\alpha_2^{-1}(A_2') \neq 0$. By (23), $\alpha_2^{-1}(A_2')$ is a direct summand of A and, by (24), $\delta^{-1}\pi_D(A_2')$ is a direct summand of A. Therefore (26) is a non–zero isomorphism between direct summands, meaning that f is pi.

2) The proof is similar to the proof of 1). Now $M = C' \oplus A_1' \oplus A_2$ implies

$$C = C' \oplus C'', \quad C'' = C \cap (A_1' \oplus A_2).$$

It then follows that

$$A = \gamma^{-1}(C') \oplus \gamma^{-1}(C''), \quad \gamma^{-1}(C') \neq 0,$$

and

$$A_2 = \alpha_2\gamma^{-1}(C') \oplus \alpha_2\gamma^{-1}(C''), \quad \alpha_2\gamma^{-1}(C') \neq 0. \tag{27}$$

By π we denote the projection from M onto $C'' \oplus D$ belonging to $M = C' \oplus C'' \oplus D$, so $\mathrm{Ker}(\pi) = C'$. By the assumption $M = C' \oplus A_1' \oplus A_2$, hence the map π induces an isomorphism of $A_1' \oplus A_2$ with $C'' \oplus D$. By (27), it follows that

$$\begin{aligned} C'' \oplus D &= \pi(A_1') \oplus \pi(A_2) \\ &= \pi(A_1') \oplus \pi\alpha_2\gamma^{-1}(C') \oplus \pi\alpha_2\gamma^{-1}(C''), \\ &\quad \pi\alpha_2\gamma^{-1}(C') \neq 0. \end{aligned} \tag{28}$$

We claim that $\pi\alpha_2\gamma^{-1}(C') \subseteq D$. For $x \in \gamma^{-1}(C')$ it follows that $\gamma(x) = (f(x), f(x) - x) \in C'$, hence $(f(-x), f(-x) + x) = -\gamma(x) \in C'$. This together with (25) implies

$$\begin{aligned} \pi\alpha_2(x) &= \pi(0, x) = \pi(f(-x), f(-x) + x) + \pi(f(x), f(x)) \\ &= \pi(f(x), f(x)) = (f(x), f(x)) \in D. \end{aligned} \tag{29}$$

Using that $\pi\alpha_2\gamma^{-1}(C') \subseteq D$ and intersecting (28) with D we get

$$D = \pi\alpha_2\gamma^{-1}(C') \oplus (D \cap L), \quad \text{where} \quad L := \pi\alpha_2\gamma^{-1}(C'') \oplus \pi(A_1'),$$

and $\pi\alpha_2\gamma^{-1}(C') \neq 0$.
Finally, we show that f induces an isomorphism between the non–zero direct summands $\gamma^{-1}(C')$ and $\delta^{-1}\pi\alpha_2\gamma^{-1}(C')$ via

$$\gamma^{-1}(C') \xrightarrow{\widehat{\alpha_2}} \alpha_2\gamma^{-1}(C') \xrightarrow{\widehat{\pi}} \pi\alpha_2\gamma^{-1}(C') \xrightarrow{\widehat{\delta^{-1}}} \delta^{-1}\pi\alpha_2\gamma^{-1}(C')$$

with isomorphisms $\widehat{\alpha_2}$, $\widehat{\pi}$, $\widehat{\delta^{-1}}$ induced by α_2, π, δ^{-1}. For $x \in \gamma^{-1}(C')$ we have by (29)

$$\delta^{-1}\pi\alpha_2(x) = \delta^{-1}(f(x), f(x)) = f(x),$$

hence

$$\gamma^{-1}(C') \ni x \mapsto f(x) \in \delta^{-1}\pi\alpha_2\gamma^{-1}(C')$$

is a non–zero isomorphism and therefore f is pi.

3) is a consequence of 2). Recall that f is a given endomorphism of A and that $C = \{(f(a), f(a) - a) \mid a \in A\}$. We note that the hypotheses are symmetrical in A_1 and A_2 and reformulate 2) using notation that does not conflict with the given f and C to obtain:

2′) *If* $g \in \mathrm{End}(A)$, $E := \{(g(a) - a, g(a)) \mid a \in A\}$, *and* $M = E' \oplus A_1 \oplus A_2'$ *for* $0 \neq E' \subseteq E$ *and* $A_2' \subseteq A_2$, *then* g *is* pi.

We apply this result with $g = 1_A - f$. Then $E = \{(a - f(a) - a, a - f(a)) \mid a \in A\} = \{(-f(a), -a) \mid a \in A\} = C$. Hence letting $E' = C'$ we have the hypotheses of 2′) and conclude that $g = 1_A - f$ is pi. □

The value of Lemma 6.10, which looks somewhat complicated, can now be seen in the proof of the main theorem.

Proof of Theorem 6.7. 1) Since always $\mathrm{Rad}(S) \subseteq \mathrm{Tot}(S)$ we only have to show that $\mathrm{Tot}(S) \subseteq \mathrm{Rad}(S)$. If $f \in \mathrm{Tot}(S)$, then also $fS \subseteq \mathrm{Tot}(S)$ as $\mathrm{Tot}(S)$ is closed under multiplication. We will show that $1 - f$ is an automorphism of A, hence $f \,\mathrm{Rad}(S)$. To do so we use the set–up in Lemma 6.10 with f and $g := 1 - f$. Since with A, also $C \cong A$ has the 2–EP, and since $M = C \oplus D = A_1 \oplus A_2$, it follows that $M = C \oplus A_1' \oplus A_2'$, $A_1' \subseteq A_1$, $A_2' \subseteq A_2$. Since f is not pi, the situation of Lemma 6.10 cannot occur which means that $A_2' = 0$, $M = C \oplus A_1'$. Now we apply the 2–EP to A_2 and the decomposition

$$M = A_1 \oplus A_2 = C \oplus A_1'.$$

Then

$$M = A_2 \oplus C' \oplus A_1'', \quad C' \subseteq C, A_1'' \subseteq A_1'.$$

By Lemma 6.10.2) $C' = 0$ since otherwise f would be pi. Therefore

$$M = A_1'' \oplus A_2.$$

Since $M = A_1 \oplus A_2$, this implies

$$A_1 = A_1'' = A_1',$$

and we have

$$M = C \oplus A_1' = C \oplus A_1. \tag{30}$$

With this we can show that $g = 1 - f$ is an automorphism.
First: $1 - f$ *is surjective.* For $x \in A$, by (30) there exist $y, z \in A$ such that

$$(0, x) = (f(y), f(y) - y) + (z, 0), \text{ so } x = (1 - f)(-y).$$

Second: $1 - f$ *is injective.* Assume $(1 - f)(y) = 0$. Then

$$(f(y), f(y) - y) = (f(y), 0) \in C \cap A_1 = \{(0, 0)\},$$

hence $f(y) = 0$ and thence $f(y) - y = -y = 0$.

2) Indirect proof. Assume that $f, g \in \mathrm{Tot}(S)$ and $f + g \notin \mathrm{Tot}(S)$. Then $f + g$ is pi and there exist $h, e \in S$ such that

$$h(f + g) = e = e^2 \neq 0.$$

Assume first that $e = 1$. Then we have

$$hf + hg = 1_A, \text{ where } hf, hg \in \mathrm{Tot}(S). \tag{31}$$

With the assumption that A has the D2–EP, we will obtain a contradiction.

Since by (31) $hf = 1 - hg$, we can use the set–up $M = A_1 \oplus A_2 = C \oplus D$ with hf and hg in place of f and g respectively. Since A has the D2–EP and $A \cong C$, also C has the D2–EP. We apply the D2–EP of C to $M = A_1 \oplus A_2 = C \oplus D$ and get

$$M = C' \oplus A_1' \oplus A_2', \quad 0 \neq C' \subseteq C, A_1' \subseteq A_1, A_2' \subseteq A_2. \tag{32}$$

If $A_2' = A_2$, then by Lemma 6.10.2), it follows that hf is pi, and then f is pi, a contradiction. Similarly, if $A_1' = A_1$, then by Lemma 6.10.3) hg is pi, and so g is pi, again a contradiction. So suppose that in (32) both $A_2' \neq A_2$ and $A_1' \neq A_1$. Then by Lemma 6.9 there exists $0 \neq C'' \subseteq^{\oplus} C'$ such that

$$M = C'' \oplus A_1' \oplus A_2$$

and by Lemma 6.10.2) again f is pi, a contradiction.

Now consider the general situation where $h(f+g) = e = e^2 \neq 0, 1$. Let

- $\iota =$ inclusion of $e(A)$ in A,
- $\pi =$ projection of A onto $e(A)$ along $(1-e)A$.

Then

$$\iota\pi = e, \quad \pi\iota = 1_{e(A)}, \quad \pi e\iota = 1_{e(A)},$$

and $hf + hg = e$ implies that

$$\pi hf\iota + \pi hg\iota = \pi e\iota = 1_{e(A)}.$$

Since the total is closed under multiplication from either side

$$\pi hf\iota, \pi hg\iota \in \mathrm{Tot}(\mathrm{End}(e(A))).$$

Since A has the D2–EP, also $e(A)$ has the D2–EP. Together we have the situation as in (31) and can use the proof of the case $e = 1$. Thus it follows that $\pi hf\iota$ or $\pi hg\iota$ is pi, and then also f or g is pi, a contradiction either way. □

The two results of our Main Theorem 6.7 establish an interesting connection between exchange properties and the total. But they also constitute an introduction to the theory of exchange modules. For more information see [21], [22]. Here we mention without proof some further noteworthy results.

The converse of the implication 6.7.2)

$$A \text{ has the D2–EP} \Rightarrow A \text{ is a total module}$$

is also true ([22]).

The converse of the implication 6.7.1)

$$A \text{ has the 2–EP} \Rightarrow A \text{ is a radical total module}$$

is NOT true, but a radical total module has a certain exchange property that is properly situated between the 2–EP and the D2–EP. It is called B2–EP ("B" for between). Thus the following is true:

$$A \text{ has the B2–EP} \Leftrightarrow A \text{ is a radical total module.}$$

We now give examples of modules with the exchange properties 2–EP and D2–EP. This will show which modules are addressed by Theorem 6.7. A direct decomposition of a module is called an LE–decomposition if the endomorphism rings of all summands are local rings. A ring is local if the set of all non–invertible elements is a two–sided ideal. Details are contained in Chapter IV.

Every module with an LE–decomposition has the D2–EP. This includes modules with local endomorphism ring. A concrete example is given in IV.2.5. These modules are studied in Chapter IV.

Let M be an indecomposable module and $S = \mathrm{End}(M)$. Then the following conditions are equivalent.

1) S is a local ring.

2) M has the D2–EP.

3) M has the 2–EP.

4) $\mathrm{Rad}(S) = \mathrm{Tot}(S)$.

Note that $\mathbb{Z}_\mathbb{Z}$ is indecomposable but $\mathbb{Z}$ ($\cong \mathrm{End}(\mathbb{Z}_\mathbb{Z})$ is not local.

If the module M is Artinian or Noetherian, then the following statements are equivalent.

1) M has a (finite) LE–decomposition,

2) M has the D2–EP.

Finally, we mention that the locally injective and the locally projective modules have the D2–EP.

These examples show that large classes of modules have exchange properties.

Chapter IV

The total of modules with LE–decompositions

1 Local rings and the total of local rings

A decomposition of a module

$$M = \bigoplus_{i \in I} M_i \tag{1}$$

is called an **LE–decomposition** (L for local, E for endomorphism ring), if the endomorphism rings $S_i := \mathrm{End}(M_i)$, $i \in I$, are all local rings.

In order to study LE–decompositions, especially in view of their total, we have to use the properties of local rings. Although these are well known (see for example [15]), we repeat them here, with proof, in a form that is useful for our later considerations. Further, we determine the total of local rings, in particular that of local endomorphism rings.

In a ring S with $1 \in S$, an element r is called **right–invertible** or **left–invertible** or **invertible** if and only if there exists $s \in S$ such that $rs = 1$ or $sr = 1$ or $rs = sr = 1$ respectively. The ring S is called **local** if and only if the set L of all elements of S that are not right–invertible is additively closed. We have the following properties.

(Loc1) *An idempotent e in an arbitrary ring S is right–invertible $\Leftrightarrow e = 1$.*

Proof. $\Leftarrow$: Clear.

$\Rightarrow$: Assume $1 = es$, $s \in S$. Then $1 = es = e^2 s = e(es) = e1 = e$. □

(Loc2) *Let S be a local ring. If for elements $s, s_1, s_2, \ldots, s_n \in S$, $s = s_1 + \cdots + s_n$ and s is right–invertible, then at least one of the s_i, $i \in \{1, \ldots, n\}$, is right–invertible.*

Proof. By definition L is additively closed. □

(Loc3) *Let S be a local ring. If $0 \neq e \in S$ is idempotent, then $e = 1$.*

Proof. Since $1 = e + (1 - e)$, we have by (Loc2) that $e = 1$ or $1 - e = 1$. Since $e \neq 0$, it follows that $e = 1$. □

(Loc4) *Let S be a local ring. If $r, s \in S$, then $rs = 1$ if and only if $sr = 1$ if and only if $rs = sr = 1$. In other words, r is right–invertible if and only if r is left–invertible if and only if r is invertible and L is the set of all non–invertible elements.*

Proof. If $rs = 1$, then it follows that

$$(sr)(sr) = s(rs)r = s1r = sr,$$

and $r(sr)s = (rs)(rs) = 1$, hence $sr \neq 0$. We showed that sr is a non–zero idempotent, hence, by (Loc3), $sr = 1$ which is all that is needed. □

(Loc5) *In a local ring S, L is a two–sided ideal and the unique largest proper right–, left–, and two–sided ideal of S.*

Proof. Assume that $r \in L$, $s \in S$ and $rs \notin L$. Then rs is right invertible, hence there exists $t \in S$ with $rst = 1 = r(st)$, hence r is right–invertible, a contradiction. Similar argument for the left side. □

(Loc6) $L = \mathrm{Rad}(S) = \mathrm{Tot}(S)$.

Proof. For any ring S, $\mathrm{Rad}(S) \subseteq \mathrm{Tot}(S)$. We will prove that $L \subseteq \mathrm{Rad}(S)$. Let A be a right ideal of S such that $A + L = S$. Then there exist $a \in A$, $r \in L$ with $a + r = 1$. Since $r \in L$, by (Loc2) it follows that a is right–invertible, so $1 \in A$ and $A = S$. This means that $L \subseteq^\circ S_S$, so $L \subseteq \mathrm{Rad}(S)$. Together we now have

$$L \subseteq \mathrm{Rad}(S) \subseteq \mathrm{Tot}(S).$$

Assume that $r \in \mathrm{Tot}(S)$ and $r \notin L$. Then there exists $s \in S$ such that $rs = 1$. But this means that r is pi, a contradiction. □

In the following we consider R–modules M with a local endomorphism ring $S := \mathrm{End}(M)$. Since the only non–zero idempotent in S is 1 (by (Loc3)), M is directly indecomposable.

The following generalization of (Loc6) holds.

(Loc7) *Let M_1, M_2 be non–zero modules with local endomorphism rings $S_i := \mathrm{End}(M_i)$, $i = 1, 2$.*

$$\begin{aligned} \mathrm{Rad}(M_1, M_2) &= \mathrm{Tot}(M_1, M_2) \\ &= \{f \in \mathrm{Hom}_R(M_1, M_2) \mid f \neq \textit{ isomorphism}\} \end{aligned}$$

and $\mathrm{Tot}(M_1, M_2)$ *is an S_2–S_1–submodule of* $\mathrm{Hom}_R(M_1, M_2)$.

Proof. We know (Theorem II.2.4) that $\mathrm{Rad}(M_1, M_2) \subseteq \mathrm{Tot}(M_1, M_2)$. We show next that $\{f \in \mathrm{Hom}_R(M_1, M_2) \mid f \neq \text{ isomorphism}\}$ is contained in $\mathrm{Rad}(M_1, M_2)$. Consider such a map f and let $g \in \mathrm{Hom}_R(M_2, M_1)$. We will prove indirectly that gf is not an isomorphism. So assume that gf is an isomorphism. Then f must be injective and since $1_{M_1} = (gf)^{-1}gf = ((gf)^{-1}g)f$, it follows that f is partially invertible. Hence there exist $0 \neq A_1 \subseteq^{\oplus} M_1$, $A_2 \subseteq^{\oplus} M_2$ such that

$$A_1 \ni a \mapsto f(a) \in A_2$$

is a non–zero isomorphism. Since M_2 is indecomposable, A_2 must equal M_2, and therefore f is also surjective. Altogether, f is an isomorphism, a contradiction.

From (Loc6) applied to S_1 it follows that $gf \in \mathrm{Rad}(S_1)$, hence by definition of radical (Definition II.2.2) $f \in \mathrm{Rad}(M_1, M_2)$. It is left to show that $\mathrm{Tot}(M_1, M_2) \subseteq \{f \in \mathrm{Hom}_R(M_1, M_2) \mid f \neq \text{ isomorphism}\}$. Let $f \in \mathrm{Tot}(M_1, M_2)$ and assume that f is an isomorphism. Then $f^{-1}f = 1_{M_1}$, hence f is pi, a contradiction. Finally, to see that $\mathrm{Tot}(M_1, M_2)$ is an S_2–S_1–submodule, we use that this is true for $\mathrm{Rad}(M_1, M_2)$. □

Why are modules with LE–decompositions interesting? To show that they are, we mention the following examples of modules with LE–decompositions.

1) M is an injective R–module and R is a Noetherian ring.
2) M is a module of finite length.
3) M is a projective and semiperfect module.
4) M is a discrete module.
5) M is a semisimple module.

2 Partially invertible homomorphisms of LE–decompositions

In the historical development of the theory of modules with LE–decompositions, one tool was a very special notion of partially invertible homomorphism. One of our goals is to connect our general notion of partially invertible and the historical one.

If we have a decomposition

$$M = C \oplus D,$$

then we denote the corresponding projectors by e_C and e_D, and we have $1_M = e_C + e_D$.

Lemma 2.1. *Assume that* $M = \bigoplus_{i \in I} M_i$ *is an LE–decomposition and*

$$M = C \oplus D, \quad C \neq 0. \tag{2}$$

Let $\iota_k : M_k \to M$ *be the inclusion map and let* $\pi_k : M \to M_k$ *be the projection belonging to* $M = \bigoplus_{i\in I} M_i$. *If, for* $k \in I$, *the map* $\pi_k e_C \iota_k$ *is an automorphism, then*

$$M = M_k \oplus \left(C \cap \bigoplus\nolimits_{i\in I, i\neq k} M_i\right) \oplus D. \tag{3}$$

Further,

$$M = e_C(M_k) \oplus \left(\bigoplus\nolimits_{i\in I, i\neq k} M_i\right), \tag{4}$$

$$M_k \ni x \mapsto e_C(x) \in e_C(M_k) \tag{5}$$

is an isomorphism and

$$C = e_C(M_k) \oplus \left(C \cap \left(\bigoplus\nolimits_{i\in I, i\neq k} M_i\right)\right). \tag{6}$$

Proof. We first apply Lemma 0.2 with $f = \iota_k$ and $g = \pi_k e_C$. Then

$$M = \mathrm{Im}(\iota_k) \oplus \mathrm{Ker}(\pi_k e_C),$$

$\mathrm{Im}(\iota_k) = M_k$, and

$$\mathrm{Ker}(\pi_k e_C) = \mathrm{Ker}(e_C) \oplus (C \cap \mathrm{Ker}(\pi_k)) = D \oplus (C \cap \mathrm{Ker}(\pi_k))$$

which establishes (3). Next we apply Lemma 0.2 with $f = e_C \iota_k$ and $g = \pi_k$. This gives (4) and (5). The decomposition (6) follows from (4) and the Modular Law. □

Remark. If in Lemma 2.1 the map $\pi_k e_D \iota_k$ is an isomorphism rather than $\pi_k e_C \iota_k$, then (4) – (6) are satisfied with D in place of C. This case will also be needed in the following.

Theorem 2.2. *Assume that* (1) *is an LE–decomposition, and*

$$M = C \oplus D, \quad C \neq 0.$$

Then there exist a $k \in I$ *and a decomposition* $C = C_1 \oplus C_2$ *with* $C_1 \cong M_k$.

Proof. Let $0 \neq c \in C$ and write

$$c = m_{i_1} + \cdots + m_{i_n}, \quad 0 \neq m_{i_j} \in M_{i_j},\ j = 1, \ldots, n. \tag{7}$$

We now identify i_1 with k in Lemma 2.1. If $\pi_{i_1} e_C \iota_{i_1}$ is an automorphism, then we have the result by (5) and (6). If $\pi_{i_1} e_C \iota_{i_1}$ is not an automorphism, then it lies in $\mathrm{Rad}(\mathrm{End}(M_{i_1}))$ since $\mathrm{End}(M_{i_1})$ is local. Then $1_{M_{i_1}} - \pi_{i_1} e_C \iota_{i_1} = \pi_{i_1} e_D \iota_{i_1}$ is invertible in $\mathrm{End}(M_{i_1})$, hence an automorphism. Then by Lemma 2.1 with D in place of C, we have using (4) that

$$M = e_D(M_{i_1}) \oplus \left(\bigoplus\nolimits_{i\in I, i\neq i_1} M_i\right) \tag{8}$$

and by (5)

$$M_{i_1} \ni x \mapsto e_D(x) \in e_D(M_{i_1})$$

is an isomorphism. Since isomorphic modules have isomorphic endomorphism rings, $e_D(M_{i_1})$ with M_{i_1} has a local endomorphism ring. Then (8) is again an LE–decomposition. Now we continue in the same manner with $i_2, \dots, i_n$. Either the first case occurs and we are finished, or it follows that

$$M = e_D(M_{i_1}) \oplus \cdots \oplus e_D(M_{i_n}) \oplus \left(\bigoplus_{i \in I \setminus \{i_1, \dots, i_n\}} M_i\right) \tag{9}$$

with

$$M_{i_j} \cong e_D(M_{i_j}), \; j = 1, \dots, n. \tag{10}$$

It now follows by (7) that

$$e_D(m_{i_1}) + \cdots + e_D(m_{i_n}) = e_D(c) = 0.$$

Since (9) is a direct sum, this implies that

$$e_D(m_{i_j}) = 0, \; j = 1, \dots, n,$$

and because of the isomorphisms (10), it follows that $m_{i_j} = 0$ for $j = 1, \dots, n$, a contradiction. Therefore the first case must occur for some M_{i_j}, $j \in \{1, \dots, n\}$. □

Suppose that besides M with LE–decomposition (1)

$$M = \bigoplus_{i \in I} M_i, \; M_i \neq 0,$$

there is another module N with an LE–decomposition

$$N = \bigoplus_{j \in J} N_j, \; N_j \neq 0. \tag{11}$$

For this decomposition we use again the notations ι_j, π_j, e_j as for (1) but with index j in place of i. Note that there is a danger of confusion.

Theorem 2.3. *Assume that* (1) *and* (11) *are LE–decompositions and that* $f \in \mathrm{Hom}_R(M, N)$. *Then* f *is partially invertible if and only if there exist* $i \in I$, $j \in J$ *such that*

$$\pi_j f \iota_i : M_i \to N_j$$

is an isomorphism.

Proof. If $\pi_j f \iota_i$ is an isomorphism, then it is pi and by Chapter II, Lemma 1.9, f is pi.

Assume, conversely, that f is pi. Then, by Chapter II, Lemma 1.1.4) there exist decompositions

$$M = C \oplus D, \; C \neq 0, \quad N = U \oplus V$$

such that

$$C \ni x \mapsto f(x) \in U \tag{12}$$

is an isomorphism. Now we apply Theorem 2.2 to get

$$C = C_1 \oplus C_2 \quad \text{with} \quad M_k \cong C_1 \neq 0. \tag{13}$$

The isomorphism (12) produces the decomposition

$$U = f(C_1) \oplus f(C_2)$$

and the isomorphism

$$C_1 \ni x \mapsto f(x) \in f(C_1). \tag{14}$$

Let $g : M_k \to C_1$ be an isomorphism which exists by (13). Further let $\iota : C_1 \to M$ be the inclusion and $\pi : N \to f(C_1)$ be the projection belonging to the decomposition

$$N = f(C_1) \oplus f(C_2) \oplus V.$$

Then

$$\pi f \iota g : M_k \xrightarrow{g} C_1 \xrightarrow{\iota} M \xrightarrow{f} N \xrightarrow{\pi} f(C_1)$$

is a non–zero isomorphism by (14). Denote by h the inverse isomorphism. Then

$$1_{M_k} = h\pi f \iota g. \tag{15}$$

For $0 \neq x \in M_k$, let $I_0 \subseteq I$, $J_0 \subseteq J$ such that

$$\textstyle\sum_{i\in I_0} e_i g(x) = g(x), \quad \sum_{j\in J_0} e_j f g(x) = f g(x). \tag{16}$$

(Caution: e_i belongs to (1) while e_j belongs to (11)) Define

$$t := 1_{M_k} - h\pi \textstyle\sum_{j\in J_0} \sum_{i\in I_0} e_j f e_i \iota g.$$

Then, by (15) and (16), it follows that $t(x) = 0$. Now consider

$$1_{M_k} = t + h\pi \textstyle\sum_{j\in J_0} \sum_{i\in I_0} e_j f e_i \iota g. \tag{17}$$

Since $\operatorname{End}(M_k)$ is local, there must be at least one summand in (17) that is an automorphism. Since $t(x) = 0$, this summand cannot be t. Assume that

$$h\pi e_j f e_i \iota g = h\pi \iota_j \pi_j f \iota_i \pi_i \iota g$$

is an automorphism. By Lemma 0.2 and since M_i is indecomposable, it follows that $h\pi\iota_j\pi_j f\iota_i$ is an isomorphism. Again by Lemma 0.2 and since N_j is indecomposable, it now follows that $\pi_j f \iota_i$ is an isomorphism which had to be shown. □

Corollary 2.4. *If* (1) *and* (11) *are LE–decompositions, then* $\operatorname{Tot}(M, N)$ *is additively closed and* $\operatorname{Tot}(S)$ *is an ideal of* $S := \operatorname{End}(M)$.

Proof. Indirect. Assume that $f, g \in \mathrm{Tot}(M, N)$ but $f + g$ is pi. Then, by Theorem 2.3, there exist $i \in I$, $j \in J$ such that $\pi_j(f+g)\iota_i$ is an isomorphism. Let h be the inverse isomorphism. Then

$$1_{M_i} = h\pi_j(f+g)\iota_i = h\pi_j f\iota_i + h\pi_j g\iota_i.$$

Since $\mathrm{End}(M_i)$ is local, at least one of the summands must be an isomorphism and hence pi. Suppose that $h\pi_j f\iota_i$ is pi, then by Lemma II.1.9, f must also be pi, a contradiction. □

Later in Section 4 we will determine conditions that imply the equality $\mathrm{Rad}(M, N) = \mathrm{Tot}(M, N)$ for modules with LE–decompositions.

Now we give an example for the case $\mathrm{Rad}(M, N) \subsetneqq \mathrm{Tot}(M, N)$.

Example 2.5. For a prime number p, let

$$M_{\mathbb{Z}} := \dot{\bigoplus}_{1 \leq n \in \mathbb{N}} \mathbb{Z}/p^n\mathbb{Z}, \quad S := \mathrm{End}(M). \tag{18}$$

Then (18) is an LE–decomposition, $\mathrm{Tot}(S)$ is an ideal and $\mathrm{Rad}(S)$ is strictly contained in $\mathrm{Tot}(S)$.

Proof. $\mathrm{End}(\mathbb{Z}/p^n\mathbb{Z}) \cong \mathbb{Z}/p^n\mathbb{Z}$ since this is a ring with 1. The non–invertible elements in this ring are the elements of $p\mathbb{Z}/p^n\mathbb{Z}$ and $1 + p^n\mathbb{Z}$ is the only non–zero idempotent. Therefore $\mathbb{Z}/p^n\mathbb{Z}$ is a local ring, (18) is an LE–decomposition, and

$$p\mathbb{Z}/p^n\mathbb{Z} = \mathrm{Rad}\left(\mathbb{Z}/p^n\mathbb{Z}\right) = \mathrm{Tot}\left(\mathbb{Z}/p^n\mathbb{Z}\right).$$

By Corollary 2.4 $\mathrm{Tot}(S)$ is an ideal. We will define $f \in S$ such that $f \in \mathrm{Tot}(S)$ but $f \notin \mathrm{Rad}(S)$. Write the elements of $\mathbb{Z}/p^n\mathbb{Z}$ in the form $a + p^n\mathbb{Z}$, $a \in \mathbb{Z}$, and define

$$\forall 1 \leq n \in \mathbb{N},\ f(a + p^n\mathbb{Z}) := pa + p^{n+1}\mathbb{Z}, \quad a \in \mathbb{Z}.$$

Obviously, $f \in S$. We show indirectly that $f \in \mathrm{Tot}(S)$. Assume to the contrary that f is pi. Then, by Theorem 2.3, there exist $i, j \in \mathbb{N}$ such that $\pi_j f\iota_i$ is an isomorphism. But $f(\mathbb{Z}/p^i\mathbb{Z}) \subseteq \mathbb{Z}/p^{i+1}\mathbb{Z}$, the index j must be $i+1$ but $\mathbb{Z}/p^i\mathbb{Z}$ and $\mathbb{Z}/p^{i+1}\mathbb{Z}$ are not isomorphic, having different cardinalities. We conclude that $f \in \mathrm{Tot}(S)$. To show that $f \notin \mathrm{Rad}(S)$ we again give a proof by contradiction. Assume that $f \in \mathrm{Rad}(S)$. Then $1_M - f$ is an automorphism of M. But $1 + p\mathbb{Z} \in \mathbb{Z}/p\mathbb{Z}$ is not in $\mathrm{Im}(1_M - f)$. In fact, let

$$m = (a_1 + p\mathbb{Z}) + (a_2 + p^2\mathbb{Z}) + \cdots + (a_k + p^k\mathbb{Z}), \quad a_i \in \mathbb{Z},$$

be an arbitrary element of M. Assume that $(1 - f)(m) = 1 + p\mathbb{Z}$, then

$$\begin{aligned}
(1-f)(m) &= (a_1 + p\mathbb{Z}) + (a_2 + p^2\mathbb{Z}) + \cdots + (a_k + p^k\mathbb{Z}) \\
&- (pa_1 + p^2\mathbb{Z}) - (pa_2 + p^3\mathbb{Z}) - \cdots - (pa_k + p^{k+1}\mathbb{Z}) \\
&= 1 + p\mathbb{Z}.
\end{aligned}$$

Comparing components we find

$$\begin{aligned} pa_k + p^{k+1}\mathbb{Z} &= 0 \Rightarrow p^k \mid a_k \Rightarrow a_k + p^k\mathbb{Z} = 0 \\ &\Rightarrow pa_{k-1} + p^k\mathbb{Z} = 0 \Rightarrow p^{k-1} \mid a_{k-1} \Rightarrow a_{k-1} + p^{k-1}\mathbb{Z} = 0 \\ &\dots \end{aligned}$$

Inductively we obtain $a_k+p^k\mathbb{Z} = a_{k-1}+p^{k-1}\mathbb{Z} = a_{k-2}+p^{k-2}\mathbb{Z} = \cdots = a_2+p^2\mathbb{Z} = a_1 + p^1\mathbb{Z} = 0$, hence $(1-f)(m) = 0 = 1 + p\mathbb{Z}$, the desired contradiction. □

This example shows that for LE–decompositions the equality $\mathrm{Rad}(S) = \mathrm{Tot}(S)$ does not always hold. But always very interesting subsets of $\mathrm{Tot}(S)$ are contained in $\mathrm{Rad}(S)$. To show this, we renew for the decomposition (1) the notations

- $\iota_i : M_i \to M$ = inclusion of M_i in M,
- $\pi_i : M \to M_i$ = projection of M onto M_i,
- $e_i := \iota_i\pi_i : M \to M$ = projector of M_i.

Then we have the following relations which we will use without explicit reference.

- $1_i := 1_{M_i} = \pi_i\iota_i$,
- $e_i\iota_i = \iota_i\pi_i\iota_i = \iota_i$,
- $\pi_i e_i = \pi_i\iota_i\pi_i = \pi_i$,
- $(1_M - e_i)\iota_i = \pi_i(1_m - e_i) = 0$.

Theorem 2.6. *Assume that* (1) *is an LE–decomposition and* $S := \mathrm{End}(M)$. *Then it follows that for all* $i \in I$,

$$e_i \,\mathrm{Tot}(S) + \mathrm{Tot}(S)e_i \subseteq \mathrm{Rad}(S) \subseteq \mathrm{Tot}(S).$$

Proof. It is enough to show that for every $f \in \mathrm{Tot}(S)$, the maps $1-e_if$ and $1-fe_i$ are invertible in S. Write

$$1 - e_if = 1 - e_i(e_if).$$

Then we see by Chapter II, Lemma 2.1.2), that $1 - e_if$ is invertible if and only if $1 - e_ife_i$ is invertible, and the same is true for $1 - fe_i$. Therefore we show that $1-e_ife_i$ is invertible. It is clear that e_iSe_i is a ring with e_i as multiplicative identity. Since $f \in \mathrm{Tot}(S)$, also $\pi_if\iota_i \in \mathrm{Tot}(S_i)$ where $S_i = \mathrm{End}(M_i)$. By (Loc6) $\mathrm{Rad}(S_i) = \mathrm{Tot}(S_i)$ and with the isomorphism III(12) it follows that $e_ife_i \in \mathrm{Rad}(e_iSe_i)$ hence $e_i - e_ife_i$ is invertible. Therefore there exists an element $e_ige_i \in e_iSe_i$ such that

$$(e_i - e_ife_i)e_ige_i = e_ige_i(e_i - e_ife_i) = e_i.$$

This implies

$$\begin{aligned} & (1_M - e_ife_i)(1_M - e_i + e_ige_i) \\ = {} & ((1_M - e_i) + (e_i - e_ife_i))(1_M - e_i + e_ige_i) \\ = {} & (1_M - e_i) + e_i = 1_M, \end{aligned}$$

and also

$$
\begin{aligned}
& (1_M - e_i + e_i g e_i)(1_M - e_i f e_i) \\
= \; & ((1_M - e_i) + e_i g e_i)((1_M - e_i) + e_i - e_i f e_i) \\
= \; & (1_M - e_i) + e_i = 1_M .
\end{aligned}
$$

This shows that $1_M - e_i f e_i$ and hence also $1_M - e_i f$ and $1_M - f e_i$ are invertible, as needed. □

Corollary 2.7. *If* $M = M_1 \oplus \cdots \oplus M_k$ *is a finite LE–decomposition and if* $S := \mathrm{End}(M)$, *then* $\mathrm{Rad}(S) = \mathrm{Tot}(S)$.

Proof. Since $1 = e_1 + \cdots + e_k$, it follows from Theorem 2.6 that

$$1_M \operatorname{Tot}(S) = e_1 \operatorname{Tot}(S) + \cdots + e_k \operatorname{Tot}(S) \subseteq \operatorname{Rad}(S).$$

Since in general $\mathrm{Rad}(S) \subseteq \mathrm{Tot}(S)$, equality follows. □

Later we consider infinite decompositions for which $\mathrm{Rad}(S) = \mathrm{Tot}(S)$. This is equivalent with several interesting phenomena.

3 The endomorphism ring of an LE–decomposition modulo its total

Again let

$$M = \bigoplus\nolimits_{i \in I} M_i, \quad M_i \neq 0, \tag{19}$$

be an LE–decomposition and set $S := \mathrm{End}(M)$. Then $\mathrm{Tot}(S)$ is an ideal of S by Corollary 2.4. It is interesting that the quotient $S/\mathrm{Tot}(S)$ has nice properties even when $\mathrm{Rad}(S) \neq \mathrm{Tot}(S)$. The main result is that $S/\mathrm{Tot}(S)$ is isomorphic to a product of endomorphism rings of vector spaces over division rings.

For $\lambda \in I$, define

$$I_\lambda := \{i \in I \mid M_i \cong M_\lambda\}, \quad \text{and} \quad H_\lambda = \bigoplus\nolimits_{i \in I_\lambda} M_i.$$

We call H_λ a **homogeneous component** of (19) (or of M) of type M_λ. Later we will use the fact that H_λ also has an LE–decomposition so that we can apply Theorem 2.3 to the partially invertible endomorphisms of H_λ.

Obviously λ is not unique in the definition of H_λ since every $i \in I_\lambda$ can replace λ. Hence we choose and fix a λ for every homogeneous component of (19). The set of all these λ we denote by Λ. Then

$$M = \bigoplus\nolimits_{\lambda \in \Lambda} H_\lambda .$$

We use again the customary notations

- $\iota_\lambda =$ inclusion of H_λ in M,

- π_λ = projection of M onto H_λ,
- $e_\lambda := \iota_\lambda \pi_\lambda$ = projector of H_λ.

We recall the definition of a direct product of rings. Let $\{R_i \mid i \in I\}$ be a family of rings. Then the product ring $\prod_{i\in I} R_i$ is the Cartesian product of the sets R_i with component–wise addition and multiplication, i.e., for $(r_i), (s_i) \in \prod_{i\in I} R_i$,

$$(r_i) + (s_i) = (r_i + s_i), \quad (r_i)(s_i) = (r_i s_i).$$

In a product ring an element is invertible or idempotent if this is true for each of the components, respectively.

For the sake of brevity we set

$$S := \operatorname{End}(M), \quad \overline{S} := S/\operatorname{Tot}(S),$$

$$S_\lambda := \operatorname{End}(H_\lambda), \quad \overline{S_\lambda} := S_\lambda/\operatorname{Tot}(S_\lambda).$$

Before starting with proofs, we survey the whole story. The fundamental ideas are easy to understand but the details of the proof are a bit long. We intend to show that $\overline{S}$ is isomorphic to a ring direct product of endomorphism rings of vector spaces over division rings. This will be done in two steps. The first step is contained in the following lemma.

Lemma 3.1 (First Isomorphism). *The following mapping is a ring isomorphism:*

$$\rho : \overline{S} \ni \overline{f} \mapsto \left(\overline{\pi_\lambda f \iota_\lambda} \mid \lambda \in \Lambda\right) \in \textstyle\prod_{\lambda\in\Lambda} \overline{S_\lambda}. \tag{20}$$

The proof will be given later.

In the second step we show that $\overline{S_\lambda}$ is isomorphic to the endomorphism ring of a vector space. To do so, we can and will assume that (19) has only one homogeneous component which means that all M_i, $i \in I$, are isomorphic to each other. Consider the fixed $\lambda \in I$ and for each $i \in I$ choose an isomorphism

$$h_i : M_i \to M_\lambda, \quad h_\lambda = 1_\lambda.$$

Then set

$$D := \operatorname{End}(M_\lambda)/\operatorname{Tot}(\operatorname{End}(M_\lambda)) = S_\lambda/\operatorname{Tot}(S_\lambda) = \overline{S_\lambda}.$$

Since S_λ is local, D is a division ring. Next we define a right vector space over D, $V := D^{(I)}$, and let $\{b_i \mid i \in I\}$ be the natural basis of V, i.e., the ith component of b_i is 1 and all other components of b_i are 0. For every M_i we then have the subspace $b_i D$ of V. For $f \in S$ we define $\hat{f} \in \operatorname{End}_D(V)$ by setting

$$\hat{f}(b_i) := \begin{cases} \displaystyle\sum_{j\in I} b_j \overline{h_j \pi_j f \iota_i h_i^{-1}} & \text{if } \pi_j f \iota_i \text{ is an isomorphism,} \\ 0 & \text{if no } \pi_j f \iota_i \text{ is an isomorphism.} \end{cases} \tag{21}$$

Notice that in this definition

$$\overline{h_j \pi_j f \iota_i h_i^{-1}} \in D = S_\lambda / \operatorname{Tot}(S_\lambda).$$

Further, since h_j is an isomorphism, the map $\pi_j f \iota_i$ is an isomorphism if and only if $h_j \pi_j f \iota_i h_i^{-1}$ is an isomorphism. Finally, one has to recognize that the sum on the right side of (21) is finite since, for $x \in M_i$, $x \neq 0$, we have $\pi_j f(x) \neq 0$ only if $j \in \operatorname{spt}(f(x))$ which is finite.

Lemma 3.2 (Second Isomorphism). *Assume that* (21) *is homogeneous. The mapping*

$$\sigma : \overline{S} = S / \operatorname{Tot}(S) \ni \overline{f} \to \hat{f} \in \operatorname{End}_D(V),$$

where $\hat{f}$ is defined by (19), *is a ring isomorphism.*

The proof will be given later. If we combine both isomorphisms we have our main result.

Main Theorem 3.3. *If M has an LE–decomposition and $S := \operatorname{End}(M)$, then $S/\operatorname{Tot}(S)$ is isomorphic to a product of endomorphism rings of vector spaces over division rings. The endomorphism rings of vector spaces correspond bijectively with the homogeneous components of M.*

Remark. The Main Theorem is the end result of a longer development in which several authors were involved. A complete presentation was given in lecture notes by F. Kasch ([16]) to which also A. Zöllner contributed greatly. In 1997 J. Zelmanowitz published a paper with the title "On the endomorphism ring of a discrete module: A theorem of F. Kasch" ([28]). The main difference between the case considered by Zelmanowitz and our case is that for a discrete module M it is true that $\operatorname{Rad}(S) = \operatorname{Tot}(S)$ while in our theorem modules with $\operatorname{Rad}(S) \neq \operatorname{Tot}(S)$ are included. In the context of LE–decompositions we recognize that the quotient $S/\operatorname{Tot}(S)$ is of importance and that the total is the appropriate concept.

In the next section we will study the consequences of $\operatorname{Rad}(S) = \operatorname{Tot}(S)$.

We now come to the proofs of the two fundamental isomorphisms. They require patience as they are a bit long.

Proof of Lemma 3.1. The additive rule $\rho(\overline{f}+\overline{g}) = \rho(\overline{f}) + \rho(\overline{g})$ follows immediately from the identity

$$\pi_\lambda (f+g)\iota_\lambda = \pi_\lambda f \iota_\lambda + \pi_\lambda g \iota_\lambda.$$

The product rule

$$\rho(\overline{f}\overline{g}) = \rho(\overline{f})\rho(\overline{g}) \tag{22}$$

requires more work. We will show that

$$\pi_\lambda (fg)\iota_\lambda - \pi_\lambda f \iota_\lambda \pi_\lambda g \iota_\lambda = \pi_\lambda (fg - f e_\lambda g)\iota_\lambda = \pi_\lambda f(1_M - e_\lambda) g \iota_\lambda \in \operatorname{Tot}(S_\lambda),$$

which implies (22). Since the total is a semi–ideal, it is enough to show that one factor of $\pi_\lambda f(1_M - e_\lambda)g\iota_\lambda$ is in the total. We show this for

$$\pi_\lambda f(1_M - e_\lambda) : M \to H_\lambda.$$

For $j \in I_\lambda$ we denote by

- ι_j^λ the inclusion of M_j in H_λ,
- π_j^λ the projection of H_λ onto M_j.

Then, for $i, j \in I_\lambda$, we have

$$\pi_j^\lambda \pi_\lambda = \pi_j, \quad \iota_\lambda \iota_j^\lambda = \iota_j, \quad e_\lambda \iota_i = \iota_\lambda \pi_\lambda \iota_i = \iota_i. \tag{23}$$

By way of contradiction, we assume that $\pi_\lambda f(1_M - e_\lambda)$ is pi. Then, by Theorem 2.3, there must exist $i \in I$, $j \in I_\lambda$ such that

$$\pi_j^\lambda \pi_\lambda f(1_M - e_\lambda)\iota_i \tag{24}$$

is an isomorphism. We consider two cases.

Case 1: $i \in I_\lambda$. Then by (23)

$$(1_M - e_\lambda)\iota_i = \iota_i - \iota_i = 0$$

which is not possible for a non–zero isomorphism.

Case 2: $i \notin I_\lambda$. Then i and j belong to different homogeneous components and then M_i and M_j are not isomorphic. Hence (24) cannot be an isomorphism.

We have shown that (24) is in the total and (22) is established.

Injectivity of ρ: If for $f \in S$ and all λ, $\pi_\lambda f \iota_\lambda \in \mathrm{Tot}(S_\lambda)$, then for all $i, j \in I_\lambda$, by (23)

$$\pi_j^\lambda \pi_\lambda f \iota_\lambda \iota_i^\lambda = \pi_j f \iota_i \in \mathrm{Tot}(M_i, M_j),$$

hence $\pi_j f \iota_i$ is not an isomorphism. But if i and j belong to different homogeneous components, then $\pi_j f \iota_i$ is not an isomorphism. These observations imply, by Theorem 2.3, that $f \in \mathrm{Tot}(S)$ which we had to show.

Surjectivity of ρ: For every $\lambda \in \Lambda$ let $f_\lambda \in S_\lambda$ be given. Define $f \in S$ by setting

$$f(x) = f_\lambda(x) \quad \text{for } x \in H_\lambda.$$

Since M is the direct sum of the H_λ, this produces a map that is well–defined on all of M. It follows that

$$\pi_\lambda f \iota_\lambda = f_\lambda,$$

and this implies that $\rho(\overline{f}) = \left(\overline{f_\lambda} \mid \lambda \in \Lambda\right)$, hence ρ is surjective and Lemma 3.1 is proved. □

Proof of Lemma 3.2. We first check that σ is a well–defined map, i.e., independent of the choice of the representative f of $\overline{f}$. If $g \in \mathrm{Tot}(S)$, then $h_j\pi_j g\iota_i h_i^{-1} \in \mathrm{Tot}(S)$ and this implies

$$\overline{h_j\pi_j(f+g)\iota_i h_i^{-1}} = \overline{h_j\pi_j f\iota_i h_i^{-1}} + \overline{h_j\pi_j g\iota_i h_i^{-1}} = \overline{h_j\pi_j f\iota_i h_i^{-1}}, \tag{25}$$

what we had to show. For arbitrary $g \in S$, it also follows from (25) that $\sigma(\overline{f+g}) = \sigma(\overline{f}) + \sigma(\overline{g})$.

We now prove $\sigma(\overline{fg}) = \sigma(\overline{f})\sigma(\overline{g})$. For this we compare the effect on the basis elements b_i, $i \in I$.

$$\widehat{fg}(b_i) = \sum_{\substack{k \in K \\ \pi_k fg\iota_i \text{ iso}}} b_k \; \overline{h_k\pi_k(fg)\iota_i h_i^{-1}} \tag{26}$$

(by "$\pi_k fg\iota_i$ iso" we mean that $\pi_k fg\iota_i$ is an isomorphism, see (21), with analogous agreements below) and

$$\left\{\begin{aligned} \hat{f}(\hat{g}(b_i)) &= \sum_{\substack{j \in I \\ \pi_j g\iota_i \text{ iso}}} \hat{f}(b_j)\; \overline{h_j\pi_j g\iota_i h_i^{-1}} \\ &= \sum_{\substack{j \in I \\ \pi_j g\iota_i \text{ iso}}} \; \sum_{\substack{k \in I \\ \pi_k f\iota_j \text{ iso}}} b_k \left(\overline{h_k\pi_k f\iota_j h_j^{-1}}\right)\left(\overline{h_j\pi_j g\iota_i h_i^{-1}}\right). \end{aligned}\right. \tag{27}$$

In order to compare the coefficients of b_k in (26) and (27), we set for fixed $i, k \in I$,

$$\alpha := \pi_k(fg)\iota_i - \textstyle\sum_{j\in J_0}(\pi_k f\iota_j)(\pi_j g\iota_i)$$

with

$$J_0 := \{j \in J \mid \pi_k f\iota_j \text{ and } \pi_j g\iota_i \text{ are both isomorphisms}\}.$$

We will show that

$$\alpha \in \mathrm{Tot}(M_i, M_k). \tag{28}$$

Then the coefficients of b_k in (26) and (27) are the same. To prove (28) we first consider the case $g(M_i) = 0$. Then $\alpha = 0 \in \mathrm{Tot}(M_i, M_k)$. Assume then that there exists $a \in M_i$ with $g(a) \neq 0$. The support of $g(a)$, $\mathrm{spt}(g(a))$, is a finite subset of I. If $j \in I$ and $j \notin \mathrm{spt}(g(a))$, then

$$(\pi_j g\iota_i)(a) = \pi_j(g(a)) = 0,$$

hence $\pi_j g\iota_i)$ cannot be an isomorphism and therefore $j \notin J_0$. This implies that $J_0 \subseteq \mathrm{spt}(g(a))$. Set

$$\beta := -\sum_{j\in \mathrm{spt}(g(a))\setminus J_0} (\pi_k f\iota_j)(\pi_j g\iota_i).$$

Then it follows, using $e_j = \iota_j \pi_j$, that

$$\begin{aligned} (\alpha+\beta)(a) &= \alpha(a)+\beta(a) = \pi_k f(g(a)) - \pi_k f\left(\sum_{j\in \operatorname{spt}(g(a))} e_j g(a)\right) \\ &= \pi_k f(g(a)) - \pi_k f(g(a)) = 0. \end{aligned}$$

Therefore $\alpha+\beta$ is not an isomorphism, hence $\alpha+\beta \in \operatorname{Tot}(M_i, M_k)$. If $j \notin J_0$, then either $(\pi_k f \iota_j)$ or $(\pi_j g \iota_i)$ is not an isomorphism; either way $(\pi_k f \iota_j)(\pi_j g \iota_i) \in \operatorname{Tot}(M_i, M_k)$ and then also $\beta \in \operatorname{Tot}(M_i, M_k)$. Since $\operatorname{Tot}(M_i, M_k)$ is additively closed, we get (28) because $\alpha = (\alpha+\beta)-\beta \in \operatorname{Tot}(M_i, M_k)$. As mentioned before, with this we have

$$\sigma(\overline{fg}) = \widehat{fg} = \hat{f}\hat{g} = \sigma(\overline{f})\sigma(\overline{g}).$$

σ is a monomorphism. By definition (21) $\hat{f} = 0$ if and only if for all $i, j \in I$, the map $\pi_j f \iota_i$ is not an isomorphism. By Theorem 2.3 it follows that $f \in \operatorname{Tot}(S)$, hence $\overline{f} = 0$.

σ is an epimorphism. Let $\varphi \in \operatorname{End}_D(V)$. If $\varphi = 0$, then for $0 = f \in S$, it follows that $\sigma(\overline{f}) = 0 = \varphi$. Assume now that $\varphi \neq 0$. For every $i \in I$, consider the finite set $K_i := \operatorname{spt}(\varphi(b_i))$ that may also be empty. If $K_i \neq \emptyset$, then there exist $0 \neq \delta_{ji} \in D$, $j \in K_i$. This means

$$\varphi(b_i) = \sum_{j\in K_i} b_j \delta_{ij}.$$

Choose $d_{ij} \in S_\lambda$ with $\overline{d_{ij}} = \delta_{ij} \in D (= S_\lambda / \operatorname{Tot}(S_\lambda))$. Since $\delta_{ij} \neq 0$, the d_{ij} must be automorphisms of M_λ. Further let

$$c_{ji} := h_j^{-1} d_{ji} h_i \in \operatorname{Hom}_R(M_i, M_j).$$

We obtain a map $f \in S$ by defining it on every M_i. For $x \in M_i$, let

$$f(x) := \begin{cases} \sum_{j\in K_i} c_{ji}(x) & \text{if} \quad K_i \neq \emptyset \\ 0 & \text{if} \quad K_i = \emptyset. \end{cases}$$

For this f we have to check $\hat{f}$ defined in (21). In the definition there appears the map

$$h_j \pi_j f \iota_i h_i^{-1} : M_\lambda \to M_\lambda.$$

If $y \in M_\lambda$, then $x := h_i^{-1}(y) \in M_i$. Now it follows if $j \in I$ and $\pi_j f \iota_i$ is an isomorphism that

$$\begin{aligned} h_j \pi_j f \iota_i h_i^{-1}(y) &= h_j \pi_j f(x) = h_j \pi_j \sum_{k\in K_i} c_{ki}(x) \\ &= h_j \pi_j \sum_{k\in K_i} h_k d_{ki} h_i(x) = h_j h_j^{-1} d_{ji}(y) = d_{ji}(y). \end{aligned}$$

This implies

$$\hat{f}(b_i) = \sum_{j\in I,\, d_{ji}\neq 0} b_j \overline{d_{ji}} = \sum_{j\in I,\, \delta_{ji}\neq 0} b_j \delta_{ji}. \qquad \square$$

In Lemma 3.2 we assumed that (21) has only one homogeneous component. In the general case of (21), the isomorphism σ can be applied to each homogeneous component H_λ, $\lambda \in \Lambda$ as follows. For H_λ, we use the notations D_λ, $V_{D_\lambda}^{(\lambda)}$, and σ_λ. Then take

$$\sigma := \prod_{\lambda\in\Lambda} \sigma_\lambda \text{ and } \rho \text{ from (20)}$$

and, with these, it follows that

$$\sigma\rho : S/\operatorname{Tot}(S) \to \prod_{\lambda\in\Lambda} \overline{S_\lambda} \to \prod_{\lambda\in\Lambda} \operatorname{End}\left(V_{D_\lambda}^{(\lambda)}\right)$$

is the isomorphism of the main result 3.3.

In the next section we will study the consequences of the equality $\operatorname{Rad}(S) = \operatorname{Tot}(S)$. This is equivalent to several interesting conditions. We will require a special case of σ which we now consider. Assume first that (19) is homogeneous and let $I_0 \subseteq I$. Then denote by e the projector of the decomposition

$$M = \left(\bigoplus_{i\in I_0} M_i\right) \oplus \left(\bigoplus_{i\in I\setminus I_0} M_i\right) \tag{29}$$

onto the first summand. To find $\hat{e}(b_i)$ we have to consider $\overline{h_j\pi_j e\iota_i h_i^{-1}}$ in (21). It is obvious that

$$h_j\pi_j e\iota_i h_i^{-1} = \begin{cases} 0 & \text{for} \quad i \neq j, i, j \in I \\ 1_D & \text{for} \quad i = j \in I_0 \\ 0 & \text{for} \quad i = j \notin I_0. \end{cases}$$

This implies

$$\hat{e}(b_i) = \begin{cases} b_i & \text{for} \quad i \in I_0 \\ 0 & \text{for} \quad i \notin I_0 \end{cases}$$

and thus $\hat{e}$ is the projector of the decomposition

$$V = \left(\bigoplus_{i\in I_0} b_i D\right) \oplus \left(\bigoplus_{i\in I\setminus I_0} b_i D\right) \tag{30}$$

onto the first summand. If (19) has several homogeneous components, then the above is true for the restriction $\pi_\lambda e\iota_\lambda$ of e to H_λ, $\lambda \in \Lambda$.

4 LE–decompositions with "very good" properties

The possibly unfamiliar notations that occur in the following main theorem will all be explained in the proof.

Main Theorem 4.1. *Let $M = \bigoplus_{i\in I} M_i$ be an LE–decomposition and $S := \operatorname{End}(M)$. Then the following statements are equivalent.*

1) $\operatorname{Rad}(S) = \operatorname{Tot}(S)$.

2) *Any LE–decomposition of M complements direct summands.*

3) *Any two LE–decompositions of M have the "replacement property".*

4) *The family $\{M_i \mid i \in I\}$ of an LE–decomposition $M = \bigoplus_{i\in I} M_i$ is a local semi–t–nilpotent family.*

The proof will be given by the chain of implications

$$1) \overset{(35)}{\Rightarrow} 2) \overset{(40)}{\Rightarrow} 3) \overset{(47)}{\Rightarrow} 4) \overset{(60)}{\Rightarrow} 1).$$

We formulate some parts of the proof as lemmas and start with a lemma that is needed for 1) $\Rightarrow$ 2).

Lemma 4.2. *Let M be an arbitrary module and set $S := \mathrm{End}(M)$.*

1) *If $S = A \oplus B$ is a decomposition of S as a right S–module, then*

$$M = A(M) \oplus B(M). \tag{31}$$

2) *If*

$$M = d(M) \oplus e(M) \tag{32}$$

with idempotents $d, e \in S$, then

$$S = dS \oplus eS \tag{33}$$

is a decomposition of S as a right S–module.

Proof. 1) Write $1 = d + e$ with $d \in A$, $e \in B$. Then for $a \in A$ we have $a = da + ea = a + 0$ and by uniqueness of the decomposition $da = a$ and $ea = 0$. The same argument applies to B and we have

$$\forall\, a \in A, da = a, ea = 0, \quad \forall\, b \in B, db = 0, eb = b. \tag{34}$$

Now, $S = A \oplus B$ implies $SM = M = A(M) + B(M)$. Assume that $ax = by$ for $a \in A$, $b \in B$, $x, y \in M$. Then it follows from (34) that $ax = dax = dby = 0$, hence $A(M) \cap B(M) = 0$ and (31) is established.

2) First we show that $dS \cap eS = 0$. Assume that for $f, g \in S$, $df = eg \in dS \cap eS$. Then for every $m \in M$,

$$df(m) = eg(m) \in d(M) \cap e(M) = 0,$$

hence $df = eg = 0$ and $dS \cap eS = 0$. To prove that $S = dS + eS$ we apply $1 - d$ to (32) using $(1 - d)d = 0$. It follows that

$$(1 - d)(M) = (1 - d)e(M).$$

With this we get

$$M = d(M) \oplus (1 - d)(M) = d(M) \oplus (1 - d)e(M).$$

The elements of M can therefore be written in the form

$$m = d(x) + (1-d)e(y), \quad x, y \in M.$$

Further, denote by e_1 the projector onto $e(M)$ belonging to (32); thus $\mathrm{Ker}(e_1) = d(M)$. We now apply the endomorphism $d + (1-d)ee_1$ to the element $m = d(x) + e(y)$, $x, y \in M$ and obtain

$$(d + (1-d)ee_1)(d(x) + e(y)) = d(x) + de(y) + (1-d)e(y) = d(x) + e(y) = m,$$

hence

$$1 = d + (1-d)ee_1 = d(1-ee_1) + ee_1.$$

This implies

$$S = d(1-ee_1)(S) + ee_1(S) \subseteq dS + eS \subseteq S,$$

hence $S = dS + eS$ and together with $dS \cap eS = 0$ we have $S = dS \oplus eS$. □

We now prove

$$1) \Rightarrow 2) \text{ of the main theorem.} \tag{35}$$

Property 2) of the main theorem ("complementing direct summands") says the following.

A direct decomposition $M = \bigoplus_{i \in I} M_i$ **complements direct summands** if for every direct summand A of M there is a set of indices $J \subseteq I$ such that $M = A \oplus \left(\bigoplus_{j \in J} M_j\right)$.

This is obviously a generalization of the replacement theorem of Steinitz, originally formulated and proved for vector spaces. It was earlier generalized to semisimple modules (see e.g., [15, 8.1.2]).

The general idea of our proof is as follows. We consider first the homogeneous case (M has only one homogeneous component). We then pass down from M to the right D–vector space V (see Section 3) and there apply the theorem of Steinitz. Then the decomposition of V has to be lifted to M. For this lifting, the assumption $\mathrm{Rad}(S) = \mathrm{Tot}(S)$ is needed. Further, we have to switch from the module to its endomorphism ring and back with the help of Lemma 4.2.

We come to the details. Assume that the decomposition of M is homogeneous. The projector of $M = A \oplus B$ onto A we denote by d. Since ρ (of Lemma 3.1) and σ (of Lemma 3.2) are ring isomorphisms, and $d^2 = d$, also $\hat{d}$ is an idempotent in $\hat{S} = \mathrm{End}_D(V)$ (see Section 3). It follows that

$$V = \hat{d}(V) \oplus (1-\hat{d})(V).$$

Since the right D–vector space V has the basis $\{b_i \mid i \in I\}$, there exists $J \subseteq I$ such that

$$V = \hat{d}(V) \oplus \left(\bigoplus_{j \in J} b_j D\right). \tag{36}$$

The statement 2) is proved, if we can lift this decomposition to the decomposition

$$M = d(M) \oplus \left(\bigoplus_{j\in J} M_j\right) = A \oplus \left(\bigoplus_{j\in J} M_j\right). \tag{37}$$

Let e be the projector onto the second summand of the decomposition

$$M = \left(\bigoplus_{i\in I\setminus J} M_i\right) \oplus \left(\bigoplus_{j\in J} M_j\right). \tag{38}$$

Then (see the end of Section 3) $\hat{e}$ is a projector onto the second summand of the decomposition

$$V = \left(\bigoplus_{i\in I\setminus J} b_i D\right) \oplus \left(\bigoplus_{j\in J} b_j D\right).$$

By (36) it follows that

$$V = \hat{d}(V) \oplus \hat{e}(V)$$

and Lemma 4.2.2) implies that

$$\hat{S} = \operatorname{End}_D(V) = \hat{d}\hat{S} \oplus \hat{e}\hat{S}.$$

Applying σ^{-1} we get

$$\overline{S} = \overline{dS} \oplus \overline{eS}. \tag{39}$$

Now we use the assumption 1) and that $\operatorname{Rad}(S)$ is small in S, and lift this decomposition modulo $\operatorname{Rad}(S) = \operatorname{Tot}(S)$ to

$$S = dS + eS + \operatorname{Rad}(S) = dS + eS.$$

Now we show that $dS \cap eS = 0$. From $dS \oplus (1-d)S$ it follows that

$$(1-d)S \cong (dS \oplus (1-d)S)/dS = (dS + eS)/dS \cong eS/(dS \cap eS).$$

Here the last isomorphism is the first isomorphism theorem (Theorem 0.1). Since $(1-d)S$ is a direct summand of S_S, it is a projective S–module, and hence also $eS/(dS \cap eS)$ is a projective S–module. Therefore

$$eS \ni es \mapsto \overline{es} \in eS/(dS \oplus eS)$$

is a split epimorphism which means that $dS \cap eS$ is a direct summand of eS. Since $S = eS \oplus (1-e)S$, the submodule $dS \cap eS$ is also a direct summand of S_S. Since, by (39), $\overline{dS} \cap \overline{eS} = 0$ and $\overline{S} = S/\operatorname{Rad}(S)$, this implies that $dS \cap eS \subseteq \operatorname{Rad}(S)$. But a direct summand that is small must be 0. Therefore we have $S = dS \oplus eS$. Now apply Lemma 4.2.2) and (37) follows. This proves 1) $\Rightarrow$ 2) in the homogeneous case.

In the general case we use again the notations of Section 3. Let d be the projector of $M = A \oplus B$ onto A. Then the components $\widehat{d_\lambda}$ are the idempotents in the endomorphism rings $\widehat{S_\lambda}$ of the vector spaces V_λ. For each $\lambda \in \Lambda$ there exists

$J_\lambda \subseteq I_\lambda$ such that (36) is satisfied for V_λ. Letting $J := \bigcup_{\lambda\in\Lambda} J_\lambda$ and letting e be the projector onto the second summand of the decomposition (38), the proof proceeds as before. $\square$

$$\text{Proof of 2)} \Rightarrow \text{3)}. \tag{40}$$

Two LE–decompositions

$$M = \bigoplus_{i\in I} M_i = \bigoplus_{j\in J} N_j$$

satisfy the **replacement property** if

for each subset $I_0 \subseteq I$ there exists a subset $J_0 \subseteq J$ such that

$$M = \left(\bigoplus_{i\in I\setminus I_0} M_i\right) \oplus \left(\bigoplus_{j\in J_0} N_j\right).$$

Let $M = \bigoplus_{i\in I} M_i = \bigoplus_{j\in J} N_j$ be LE–decompositions and suppose that

$$M = \left(\bigoplus_{i\in I\setminus I_0} M_i\right) \oplus \left(\bigoplus_{i\in I_0} M_i\right).$$

Then, in the definition of "complementing direct summands", let $A = \bigoplus_{i\in I\setminus I_0} M_i$ and $B = \bigoplus_{i\in I_0} M_i$. We then apply 2) to $M = A \oplus B$ and the decomposition $M = \bigoplus_{j\in J} N_j$ and obtain 3). $\square$

Later in the proof of the implication 3) $\Rightarrow$ 4) we need the following result on the replacement property.

Lemma 4.3. *Let M be a module with an LE–decomposition*

$$M = \bigoplus_{i\in I} M_i \tag{41}$$

and assume that the replacement property is satisfied for any LE–decomposition of M. If $I_0 \subseteq I$, then the subsum

$$N := \bigoplus_{i\in I_0} M_i \tag{42}$$

has the replacement property for any LE–decomposition.

Proof. Let

$$N = \bigoplus_{j\in J} A_j = \bigoplus_{k\in K} B_k$$

be LE–decompositions. By hypothesis

$$M = \left(\bigoplus_{j\in J} A_j\right) \oplus \left(\bigoplus_{i\in I\setminus I_0} M_i\right) = \left(\bigoplus_{k\in K} B_k\right) \oplus \left(\bigoplus_{i\in I\setminus I_0} M_i\right).$$

Let $J_0 \subseteq J$. Then by assumption there exist modules $C_\ell \in \{B_k, M_i \mid k \in K, i \in I \setminus I_0\}$ that can replace the summands A_j, $j \in J_0$. But none of the C_ℓ can be an

M_i since these belong to the complementary summand $\bigoplus_{i\in I\setminus I_0} M_i$ of $\bigoplus_{j\in J} A_j$. Hence there is $K_0 \subseteq K$ such that

$$M = \left(\bigoplus_{j\in J\setminus J_0} A_j\right) \oplus \left(\bigoplus_{i\in I\setminus I_0} M_i\right) \oplus \left(\bigoplus_{k\in K_0} B_k\right). \tag{43}$$

Since the first and the last summand of (43) are contained in N and the middle summand in (43) has intersection 0 with N, it follows by the Modular Law that

$$N = \left(\bigoplus_{j\in J\setminus J_0} A_j\right) \oplus \left(\bigoplus_{k\in K_0} B_k\right). \qquad \square$$

5 Locally semi–t–nilpotent families

Definition 5.1. A family of LE–modules $\{M_i \mid i \in I\}$ is called **locally semi-t-nilpotent** (lstn for short) if and only if for every infinite sequence of pairwise different elements

$$i_1, i_2, i_3, \ldots \in I \tag{44}$$

and for every sequence of homomorphisms

$$f_1, f_2, f_3, \ldots \text{ with } f_j \in \mathrm{Tot}(M_{i_j}, M_{i_{j+1}}) \tag{45}$$

and for every $x \in M_{i_1}$, there exists $n \in \mathbb{N}$ such that

$$f_n f_{n-1} f_{n-2} \cdots f_2 f_1(x) = 0. \tag{46}$$

Since the M_i are LE–modules (which means that $\mathrm{End}(M_i)$ is a local ring), it is true that $\mathrm{Tot}(M_{i_j}, M_{i_{j+1}})$ is the set of homomorphisms from M_{i_j} to $M_{i_{j+1}}$ that are not isomorphisms.

We note further that for a finite set I, no sequence (44) exists. Therefore a finite family $\{M_i \mid i \in I\}$ is always lstn.

The word "locally" in the definition means that n in (46) may depend on x; "semi" means that in (44) the i_j are all different from one another; t stands for "transfinite" meaning that the n appearing in (46) are not bounded. Altogether this definition constitutes a weak finiteness condition.

In the step 3) $\Rightarrow$ 4) of the proof of Theorem 4.1 we will have to change the sequences (45) in such a way that if the new sequence satisfies (46), then the original sequence (45) satisfies (46). There are two such changes.

1. *Change.* If for any $m \in \mathbb{N}$, the sequence

$$f_m, f_{m+1}, f_{m+2}, \ldots, \quad \text{with } f_j \in \mathrm{Tot}(M_{i_j}, M_{i_{j+1}}),$$

satisfies the condition (46), then the sequence (45) satisfies (46).

In fact, given $x \in M_1$, let $y := f_{m-1}f_{m-2}\cdots f_1(x) \in M_{i_m}$. Then since the new sequence satisfies (46), there exists $n \geq m$ such that

$$f_n f_{n-1}\cdots f_m(y) = 0,$$

that is, (46) is satisfied for (45). □

2. *Change.* For any integers

$$1 \leq m_1 < m_2 < m_3 < \cdots$$

we define

$$\begin{aligned} g_1 &:= f_{m_2}f_{m_2-1}\cdots f_{m_1+1}f_{m_1} \in \mathrm{Tot}(M_{m_1}, M_{m_2}),\\ g_2 &:= f_{m_3}f_{m_3-1}\cdots f_{m_2+1}f_{m_2} \in \mathrm{Tot}(M_{m_2}, M_{m_3}),\\ g_3 &:= f_{m_4}f_{m_4-1}\cdots f_{m_3+1}f_{m_3} \in \mathrm{Tot}(M_{m_3}, M_{m_4}),\\ &\ \vdots \end{aligned}$$

If the sequence $g_1, g_2, g_3, \ldots$ satisfies the condition (46), then obviously also (45). □

For the sake of shorter notation we replace the indices $i_1, i_2, i_3, \ldots$ by $1, 2, 3, \ldots$ and then $f_j \in \mathrm{Tot}(M_j, M_{j+1})$.

We prove next the implication

$$3) \Rightarrow 4) \text{ of Theorem 4.1.} \tag{47}$$

By Lemma 4.3 we can and will apply the replacement property to decompositions

$$N := \bigoplus\nolimits_{j\in\mathbb{N}} M_j. \tag{48}$$

By using the sequence $f_1, f_2, f_3, \ldots$ we construct two LE–decompositions of N to which we apply the replacement property. Define

$$N_j := (1 - f_j)(M_j),\ j \in \mathbb{N}, \tag{49}$$

where $1 = 1_j$ is the identity on M_j. By Lemma 0.4

$$M_j \ni x \mapsto (1 - f_j)(x) \in N_j \tag{50}$$

is an isomorphism and

$$N_i \oplus M_{i+1} = M_i \oplus M_{i+1},\ i \in \mathbb{N}. \tag{51}$$

Thus M_j and N_j are both LE–modules. It is clear that $N = N_1 \oplus \left(\bigoplus_{j>1} M_j\right) = N_1 \oplus N_2 \oplus \left(\bigoplus_{j>2} M_j\right) = N_1 \oplus N_2 \oplus N_3 \oplus \left(\bigoplus_{j>3} M_j\right) = \cdots$ and that $\sum_{j\in\mathbb{N}} N_j = \bigoplus_{j\in\mathbb{N}} N_j$. Let

$$F := \bigoplus\nolimits_{j\in\mathbb{N}} N_j \subseteq N. \tag{52}$$

Now for N in (48), using (51), we have the following two decompositions.

$$N = M_1 \oplus N_2 \oplus M_3 \oplus N_4 \oplus \cdots = \left(\bigoplus_{j\in\mathbb{N}} M_{2j-1}\right) \oplus \left(\bigoplus_{j\in\mathbb{N}} N_{2j}\right) \tag{53}$$

and

$$N = N_1 \oplus M_2 \oplus N_3 \oplus M_4 \oplus \cdots = \left(\bigoplus_{j\in\mathbb{N}} N_{2j-1}\right) \oplus \left(\bigoplus_{j\in\mathbb{N}} M_{2j}\right). \tag{54}$$

Now we apply the replacement property to (53) and (54). We will substitute the M_{2j-1} in (53) by summands of (54). Thus there exist $K \subseteq \mathbb{N}$, $L \subseteq \mathbb{N}$ such that from (53) we obtain

$$N = \underbrace{\left(\bigoplus_{j\in K} N_{2j-1}\right) \oplus \left(\bigoplus_{j\in L} M_{2j}\right)}_{\text{these are the new summands}} \oplus \left(\bigoplus_{j\in\mathbb{N}} N_{2j}\right). \tag{55}$$

The proof of the implication 3) $\Rightarrow$ 4) will be completed in three steps.

Step 1: The decomposition (55) implies that the sequence

$$f_1, f_2, f_3, \ldots \text{ with } f_j \in \operatorname{Tot}(M_j, M_{j+1}) \tag{56}$$

contains infinitely many f_j that are not monomorphisms.

Proof. Indirect. Assume that (56) has only finitely many non–monomorphisms. We employ the first change. By erasing finitely many of the f_j at the beginning of (56) we obtain a chain in which all terms f_j are monomorphisms. We denote this chain again by (56). Since $M_1 \subseteq N$ and the elements of M_1 do not occur elsewhere in (55), N_1 must appear in (55), that is, $1 \in K$. Since f_{2j} is a monomorphism but not an isomorphism (recall that $f_{2j} \in \operatorname{Tot}(M_{2j}, M_{2j+1})$), it is not an epimorphism, hence $f_{2j}(M_{2j}) \subsetneqq M_{2j+1}$. Therefore N_{2j+1} must also be in (55), hence $j+1 \in K$ which implies that $K = \mathbb{N}$. With this (55) becomes

$$N := \left(\bigoplus_{j\in\mathbb{N}} N_j\right) \oplus \left(\bigoplus_{j\in L} M_{2j}\right). \tag{57}$$

To get a contradiction in this indirect proof, we have to consider L in (57). Assume that L contains more than one element and let $m < n$ be the first two integers in L. Then we arrange (57) in the form

$$\begin{aligned} N &= N_1 \oplus \cdots \oplus N_{2m-1} \oplus M_{2m} \\ &\oplus [N_{2m} \oplus \cdots \oplus N_{2n-1} \oplus M_{2n}] \oplus N_{2n} \oplus (\text{further summands}). \end{aligned}$$

We now apply (51) to the bracketed summand with $i+1 = 2n, 2n-1, \ldots, 2m+1$. It follows that

$$N = N_1 \oplus \cdots \oplus N_{2m-1} \oplus M_{2m} \oplus (M_{2m} \oplus \cdots \oplus M_{2n}) \oplus N_{2n} \oplus \cdots .$$

But this is obviously wrong, since M_{2m} cannot occur twice as a summand in a direct sum. Consequently, L contains only one element or is empty.

We check both cases and show that they are not possible. Then our indirect proof is finished.

$L = \{n\}$: Now (57) has the form

$$N := \left(\bigoplus_{j\in\mathbb{N}} N_j\right) \oplus M_{2n}. \tag{58}$$

Since f_{2n} is a monomorphism, it is not an epimorphism. Hence there exists an element $c \in M_{2n+1}$ that is not in $f_{2n}(M_{2n})$, so also $c \neq 0$. Since $c \in N$, it can be written, by (58), in the form

$$c = \sum_{j=1}^{t}(1 - f_j)(a_j) + b, \tag{59}$$

with $a_j \in M_j$, $b \in M_{2n}$, and since $c \neq 0$ we can assume that $(1 - f_t)(a_t) \neq 0$. Since $c \in M_{2n+1}$, t cannot be $\leq 2n$. Here we use the assumption that $c \notin f_{2n}(M_{2n})$. But if $t \geq 2n + 1$, then it follows from

$$(1 - f_t)(a_t) = a_t - f_t(a_t) \neq 0$$

that $a_t \neq 0$, and since f_t is a monomorphism, also $f_t(a_t) \neq 0$. But this element is the only coefficient of (59) in M_{t+1} (as $t + 1 \geq 2n + 2$). Hence the right-hand side of (59) cannot be equal to c. Thus (57) cannot be satisfied if $L = \{n\}$.

$L = \emptyset$: In this case (57) has the form $N = \bigoplus_{j\in\mathbb{N}} N_j$. We show now that $c \in M_1$, $c \neq 0$, cannot be in N. Assume to the contrary that

$$c = \sum_{j=1}^{t}(1 - f_j)(a_j),\ a_j \in M_j,\ (1 - f_t)(a_t) \neq 0.$$

Then, f_t being injective, $f_t(a_t) \neq 0$, $f_t(a_t) \in M_{t+1}$, and $t + 1 \geq 2$. Hence the right side cannot equal $c \in M_1$, a contradiction. This means that the case $L = \emptyset$ is also not possible. Altogether we have seen that the decomposition (55) is not possible if (56) has only finitely many non–monomorphisms. □

Step 2. We assume now that (56) contains infinitely many non–monomorphisms and we show that

$$N = \bigoplus_{j\in\mathbb{N}} N_j.$$

For the proof, we use the first and the second changes above. Suppose that in (56)

$$f_{m_1}, f_{m_2}, f_{m_3}, \ldots, \quad m_1 < m_2 < m_3 < \cdots$$

are not monomorphisms. Define

$$\begin{aligned} g_1 &:= f_{m_2-1}f_{m_2-2}\cdots f_{m_1+1}f_{m_1}, \\ g_2 &:= f_{m_3-1}f_{m_3-2}\cdots f_{m_2+1}f_{m_2}, \\ g_3 &:= f_{m_4-1}f_{m_4-2}\cdots f_{m_3+1}f_{m_3}, \\ &\ \ \vdots \end{aligned}$$

Then none of the g_j are monomorphisms. By Change 2. it is enough to prove that the sequence $g_1, g_2, g_3, \ldots$ has the lstn–property (see Definition 5.1). We change the notation and denote the sequence $g_1, g_2, g_3, \ldots$ again by $f_1, f_2, f_3, \ldots$ and use (55). Since f_{2j} is not monomorphic, there exists $0 \neq x \in \mathrm{Ker}(f_{2j}) \subseteq M_{2j}$. Then

$$x = (1 - f_{2j})(x) \subseteq M_{2j} \cap N_{2j},$$

a contradiction, and this would also be a contradiction if $L \neq \emptyset$ in (55). Hence $L = \emptyset$ in (55). Next we show that $K = \mathbb{N}$. Since by (55) and (52)

$$N = \left(\bigoplus\nolimits_{j \in K} N_{2j-1}\right) \oplus \left(\bigoplus\nolimits_{j \in \mathbb{N}} N_{2j}\right) \subseteq F \subseteq N$$

we have

$$N = \bigoplus\nolimits_{j \in \mathbb{N}} N_j = \bigoplus\nolimits_{j \in \mathbb{N}} M_j.$$

Step 3. Now take $0 \neq x \in M_1$. Since $x \in M_1 \subseteq N = \bigoplus_{j \in \mathbb{N}} N_j$, there exists a representation

$$x = \sum_{j=1}^{n} (1 - f_j)(a_j), \quad a_j \in M_j.$$

Arranging the summands in accordance with (48), i.e., $N = \bigoplus_{j \in \mathbb{N}} M_j$, we obtain that

$$x = a_1 + (-f_1(a_1) + a_2) + \cdots + (-f_{n-1}(a_{n-1}) + a_n) + (-f_n(a_n)).$$

Since (48) is a direct sum, this implies

$$\begin{aligned} x &= a_1, \\ f_1(a_1) &= a_2, \\ f_2(a_2) &= a_3, \\ &\vdots \\ f_{n-1}(a_{n-1}) &= a_n, \\ f_n(a_n) &= 0, \end{aligned}$$

and it follows that

$$\begin{aligned} f_n f_{n-1} f_{n-2} \cdots f_2 f_1(x) &= f_n f_{n-1} f_{n-2} \cdots f_2(a_2) \\ &= \cdots = f_n f_{n-1}(a_{n-1}) = f_n(a_n) = 0, \end{aligned}$$

which is what we had to show. This concludes the proof of 3) $\Rightarrow$ 4). □

We now come to the proof of

$$4) \Rightarrow 1). \tag{60}$$

The idea of the proof is simple. We know already that $\mathrm{Rad}(S) \subseteq \mathrm{Tot}(S)$. To show the inverse inclusion, we use Lemma II.2.1, that is we prove: If $f \in \mathrm{Tot}(S)$, then $1 - f$ is invertible in S. For the proofs the projectors e_i, $i \in I$, belonging to the LE–decomposition $M = \bigoplus_{i\in I} M_i$ are an important tool. We denote the set of the e_i by E. Then E is an orthogonal set of idempotents in S. To show that $1 - f$ is injective, we do not need the assumption 4). It is only needed later to prove that $1 - f$ is surjective.

For $x \in M$, let

$$d := \sum_{i\in \mathrm{spt}(x)} e_i,$$

where $\mathrm{spt}(x)$ is the support of x, i.e., the set of all $i \in I$ with $e_i x \neq 0$. Then d is a finite sum and an idempotent with $x = dx$. (If $x = 0$, then $\mathrm{spt}(x) = \emptyset$ and $d = 0$) It follows for $f \in S$ that

$$(1 - f)(x) = x - f(x) = x - f(dx) = (1 - fd)(x). \tag{61}$$

If $f \in \mathrm{Tot}(S)$, then it follows by Theorem 2.6 that $fd \in \mathrm{Rad}(S)$, hence $1 - fd$ is invertible. We infer that

$$x \;=\; (1 - fd)^{-1}(1 - fd)(x) \overset{(61)}{=} (1 - fd)^{-1}(1 - f)(x).$$

If $(1 - f)(x) = 0$, then this implies that $x = 0$, hence $1 - f$ is injective. To prove that $1 - f$ is surjective we require condition 4). The idea of the proof is easy but the details are complicated. Therefore, we first give an outline of the proof.

For given $f \in \mathrm{Tot}(S)$, $f \neq 0$, and $x \in M$, $x \neq 0$, we construct two sequences

$$\begin{aligned} \varphi_1, \varphi_2, \varphi_3, \ldots &\in \mathrm{Tot}(S), \\ g_1, g_2, g_3, \ldots &\in S, \end{aligned}$$

such that for every $n \in \mathbb{N}$,

$$(1 - \varphi_n \varphi_{n-1} \cdots \varphi_1)(x) = (1 - f)(g_n(x)). \tag{62}$$

In this construction we again have to use Theorem 2.6 which says that for $f \in \mathrm{Tot}(S)$ and $e \in E$ it follows that $ef \in \mathrm{Rad}(S)$, hence $1 - ef$ is invertible. We show that there exists $n \in \mathbb{N}$ with $(\varphi_n \varphi_{n-1} \cdots \varphi_1)(x) = 0$, from which it follows by (62) that

$$(1 - f)(g_n(x)) = (1 - \varphi_n \varphi_{n-1} \cdots \varphi_1)(x) = x,$$

so that $1 - f$ is surjective as desired. To be able to apply the local–semi–t–nilpotent property, where the homomorphisms are in $\mathrm{Tot}(M_i, M_j)$ with different i, j, we have to insert in the product $\varphi_n \varphi_{n-1} \cdots \varphi_1$ in (62) certain idempotents $e_{i_n}, e_{i_{n-1}}, \ldots, e_{i_1}, e_{i_0} \in E$:

$$e_{i_n} \varphi_n e_{i_{n-1}} \varphi_{n-1} e_{i_{n-2}} \cdots e_{i_1} \varphi_1 e_{i_0}.$$

Using the $(n+1)$–tuples $(i_n, i_{n-1}, \ldots, i_1, i_0)$ we define a tree. The lstn–property implies that the tree does not have an infinite branch. Then by the König Graph Theorem it follows that the tree must be finite and this implies that there exists an n with $(\varphi_n \varphi_{n-1} \cdots \varphi_1)(x) = 0$.

This concludes the outline of the proof of the surjectivity of $1 - f$. We now come to the details of the proof that are largely contained in the following two lemmas.

Lemma 5.2. *If $\varphi_1, \varphi_2, \ldots, \varphi_n \in S$ and $x \in M$, then*

$$\varphi_n \varphi_{n-1} \cdots \varphi_1(x) = \tag{63}$$

$$\sum \left\{ e_{i_n} \varphi_n e_{i_{n-1}} \varphi_{n-1} e_{i_{n-2}} \cdots e_{i_1} \varphi_1 e_{i_0}(x) \,\middle|\, \begin{array}{c} i_n, i_{n-1}, \ldots, i_0 \in I \\ e_{i_n} \varphi_n e_{i_{n-1}} \cdots e_{i_1} \varphi_1 e_{i_0}(x) \neq 0 \end{array} \right\}.$$

Proof. Induction on n.

$n = 0$.

$$x = \sum_{i_0 \in \operatorname{spt}(x)} e_{i_0}(x) = \sum_{i_0 \in I, e_{i_0}(x) \neq 0} e_{i_0}(x). \tag{64}$$

This equation is also true for $x = 0$, since then the sum is empty and hence equal to 0 by convention.

Induction step. Assume that (63) is valid and apply φ_{n+1} to obtain

$$\varphi_{n+1} \varphi_n \varphi_{n-1} \cdots \varphi_1(x) = \sum_{\text{as in (63)}} \varphi_{n+1} e_{i_n} \varphi_n e_{i_{n-1}} \varphi_{n-1} e_{i_{n-2}} \cdots e_{i_1} \varphi_1 e_{i_0}(x).$$

Now apply (64) to each summand $\varphi_{n+1} e_{i_n} \varphi_n e_{i_{n-1}} \varphi_{n-1} e_{i_{n-2}} \cdots e_{i_1} \varphi_1 e_{i_0}(x)$ in place of x and (63) follows for $n+1$. □

Set

$$K_n := \bigcup_{\text{as in (63)}} \{i_n, \ldots, i_1, i_0\}, \ n \in \mathbb{N}, \text{ in particular } K_0 := \operatorname{spt}(x). \tag{65}$$

Remark. $\operatorname{spt}(\varphi_n \varphi_{n-1} \cdots \varphi_1(x)) \subseteq K_n$.

Proof. If $e\varphi_n \varphi_{n-1} \cdots \varphi_1)(x) \neq 0$ for an idempotent $e \in E$, then by (63) e must equal one of the e_n in (63) which are all in K_n. □

The next lemma is the key to subsequent considerations.

Lemma 5.3. *If $M = \bigoplus_{i \in I} M_i$ is an LE–decomposition, I is infinite, $0 \neq f \in \operatorname{Tot}(S)$, and $0 \neq x \in M$, then there exist sequences*

$$\begin{aligned} \varphi_1, \varphi_2, \varphi_3, \ldots &\in \operatorname{Tot}(S), \\ g_1, g_2, g_3, \ldots &\in S, \end{aligned}$$

with the following properties.

$$(1-\varphi_n\varphi_{n-1}\cdots\varphi_1)(x) = (1-f)(g_n(x)). \tag{66}$$

Furthermore, for $n \in \mathbb{N}$, *and* $i_{n-1}, i_{n-2}, \ldots, i_1, i_0 \in I$,

$$\begin{aligned} &\textit{if } y := \varphi_n e_{i_{n-1}}\varphi_{n-1}e_{i_{n-2}}\cdots e_{i_1}\varphi_1 e_{i_0}(x) \neq 0,\\ &\textit{then } \operatorname{spt}(y)\cap K_{n-1} = \emptyset \textit{ with } K_{n-1} \textit{ from } (65). \end{aligned} \tag{67}$$

Proof. Step 1. Let $I_0 := \operatorname{spt}(x)$ and let $d := \sum_{i\in I_0} e_i$. Then $d(x) = x$ and $d^2 = d \neq 0$. Since $df \in \operatorname{Rad}(S)$, the element $1 - df$ is invertible. Define

$$\varphi := (1-d)f(1-df)^{-1} \in \operatorname{Tot}(S), \quad g := (1-df)^{-1}.$$

Then

$$\varphi = (1-d)fg, \quad g^{-1} = 1 - df.$$

It further follows that

$$(1-\varphi)(x) = (1-f)g(x), \tag{68}$$

because $(1-\varphi)(x) = (g^{-1}g - (1-d)fg)(x) = (g^{-1} - (1-d)f)g(x) = (1 - df - f + df)g(x) = (1-f)g(x)$. For $i \in I_0$ it follows by definition of d that

$$e_i(1-d) = e_i - e_i = 0;$$

hence $e_i\varphi = 0$. Thus for arbitrary $z \in M$ it follows that

$$\operatorname{spt}(\varphi(z)) \cap I_0 = \emptyset. \tag{69}$$

Step 2. We now verify (66) and (67) by induction on n. To do so we define inductively three sequences

$$\begin{aligned} d_1, d_2, d_3, \ldots &\in S, \quad d_i^2 = d_i,\\ \varphi_1, \varphi_2, \varphi_3, \ldots &\in \operatorname{Tot}(S), \text{ and}\\ g_1, g_2, g_3, \ldots &\in S \end{aligned}$$

such that (66) and (67) are satisfied.

Let

$$I_0 = K_0 := \operatorname{spt}(x), \quad d_1 := \textstyle\sum_{i\in K_0} e_i.$$

Then the assumptions in Step 1. of the proof are satisfied and we get

$$\varphi_1 := \varphi, \quad g_1 := g,$$

with φ and g from Step 1. Then, by (68) and (69)

$$(1-\varphi_1)(x) = (1-f)g_1(x) \quad \text{and} \quad \operatorname{spt}(\varphi_1 e_{i_0}(x)) \cap K_0 = \emptyset \quad \forall\, i_0 \in \operatorname{spt}(x) = K_0.$$

Induction step. By induction hypothesis Lemma 5.3 is satisfied for n. Then K_n is defined by (65) and

$$d_{n+1} := \sum_{i \in K_n} e_i.$$

For $i \in K_n$ it then follows that

$$e_i(1 - d_{n+1}) = e_i - e_i = 0.$$

Now apply Step 1. to the element $\varphi_n \varphi_{n-1} \cdots \varphi_1(x)$ in place of x. Accordingly there exist

$$\varphi_{n+1} := (1 - d_{n+1}) f (1 - d_{n+1} f)^{-1} \in \operatorname{Tot}(S), \quad g := (1 - d_{n+1} f)^{-1},$$

such that

$$(1 - \varphi_{n+1}) \varphi_n \varphi_{n-1} \cdots \varphi_1(x) = (1 - f) g(\varphi_n \varphi_{n-1} \cdots \varphi_1(x)).$$

This together with (66) gives

$$\begin{aligned} & (1 - \varphi_{n+1}) \varphi_n \varphi_{n-1} \cdots \varphi_1(x) + (1 - \varphi_n \varphi_{n-1} \cdots \varphi_1)(x) \\ = \; & (1 - \varphi_{n+1} \varphi_n \cdots \varphi_1)(x) \\ = \; & (1 - f) g((\varphi_n \varphi_{n-1} \cdots \varphi_1)(x)) + (1 - f) g_n(x) \\ = \; & (1 - f)(g \varphi_n \varphi_{n-1} \cdots \varphi_1 + g_n)(x). \end{aligned}$$

Define

$$g_{n+1}(x) := (g \varphi_n \varphi_{n-1} \cdots \varphi_1 + g_n)(x).$$

Then (66) is satisfied for $n+1$. Also (67) is satisfied for $n+1$ since

$$\operatorname{spt}(\varphi_{n+1} e_{i_n} \varphi_n e_{i_{n-1}} \cdots \varphi_1 e_{i_0}(x)) \cap K_n = \emptyset$$

by definition of the ϕ_{n+1} and d_{n+1}. □

The next goal is to show that there exists an n with $(\varphi_n \varphi_{n-1} \cdots \varphi_1)(x) = 0$. It then follows by (66) that

$$(1 - \varphi_n \varphi_{n-1} \cdots \varphi_1)(x) = x = (1 - f) g_n(x).$$

This shows that $1 - f$ is surjective and concludes the proof of 4) $\Rightarrow$ 1).

To achieve this goal we have to construct a "tree". A general tree is defined as follows. Let I be an arbitrary set and for every $n \in \mathbb{N}_0$ let

$$B_n \subseteq I^{n+1} = \{(i_n, i_{n-1}, \ldots, i_1, i_0) \mid i_j \in I\}.$$

A sequence

$$\mathfrak{B} := (B_0, B_1, B_2, \ldots)$$

is called a **tree**, if the following conditions (i) and (ii) are satisfied.

(i) All sets B_n are finite (or empty).

(ii) If $(i_n, i_{n-1}, \ldots, i_1, i_0) \in B_n$, then $(i_{n-1}, \ldots, i_1, i_0) \in B_{n-1}$.

We have to consider some properties of a tree. If $k \leq n$ and if $(i_n, \ldots, i_0) \in B_n$, then by (ii) it follows that $(i_k, \ldots, i_0) \in B_k$. Hence, if $B_n = \emptyset$, then $B_k = \emptyset$ for all $k \leq n$, The element $(i_n, \ldots, i_k, \ldots, i_0) \in B_n$ is called an **extension** of $(i_k, \ldots, i_0) \in B_k$, $k \leq n$. The element $(i_k, i_{k-1}, \ldots, i_0) \in B_k$ has **arbitrarily long** extensions, if for every $n \in \mathbb{N}_0$ with $n \geq k$, the set B_n contains an extension of $(i_k, \ldots, i_0) \in B_k$. If $(i_k, \ldots, i_0) \in B_k$ does not have arbitrarily long extensions, then there exists $m > k$ such that B_m does not contain an extension of $(i_k, \ldots, i_0)$. If so, in every extension B_n, $n \geq m$, there is no extension of $(i_k, \ldots, i_0)$. If all elements of B_k do not have arbitrarily long extensions, then there exists $B_n = \emptyset$ and also $B_t = \emptyset$ for $t \geq n$. A tree $\mathfrak{B}$ is called an **infinite tree** if $B_n \neq \emptyset$ for all $n \in \mathbb{N}_0$. An infinite sequence of elements of I

$$(\ldots, i_n, i_{n-1}, \ldots, i_1, i_0) \tag{70}$$

is called **a branch of infinite length** of $\mathfrak{B}$, if for every $n \in \mathbb{N}_0$ $(i_n, i_{n-1}, \ldots, i_1, i_0) \in B_n$.

We will need the König Graph Theorem which we include with proof.

König Graph Theorem 5.4. *Every infinite tree has a branch of infinite length.*

Proof. Let $\mathfrak{B} = (B_0, B_1, B_2, \ldots)$ be an infinite tree and assume that

$$(i_n, i_{n-1}, \ldots, i_1, i_0) \in B_n$$

is an element that has arbitrarily long extensions. Then there must exist an element

$$(i_{n+1}, i_n, \ldots, i_1, i_0) \in B_{n+1}$$

that has arbitrarily long extensions, as otherwise $(i_n, i_{n-1}, \ldots, i_1, i_0)$ could not have arbitrarily long extensions. This observation can be used to define (70) inductively.

$n = 0$. Write $B_0 = \{(j_1), \ldots, (j_r)\}, j_i \in I$. Then at least one of the elements (j_i) has arbitrarily long extensions. Denote one such element by (i_0). Then B_1 contains an extension of (i_0) that is an element that has arbitrarily long extensions. Let (i_1, i_0) be such an element.

Induction step. Assume that $(i_n, \ldots, i_0) \in B_n$ has been found having arbitrarily long extensions. Then there exists

$$(i_{n+1}, i_n, \ldots, i_1, i_0) \in B_{n+1}$$

having arbitrarily long extensions. Since $\mathfrak{B}$ is an infinite tree this process can be continued to obtain (70). □

We now return to our case where I is the index set of $\bigoplus_{i\in I} M_i$. With our notation we define

$$\begin{aligned} B_0 &:= \{i_0 \mid e_{i_0}x \neq 0\} = \operatorname{spt}(x), \\ B_n &:= \{(i_n, \ldots, i_0) \in I^{n+1} \mid e_{i_n}\varphi_n e_{i_{n-1}}\varphi_{n-1}\cdots e_{i_1}\varphi_{i_1}e_{i_0}(x) \neq 0\}. \end{aligned}$$

Then the conditions (i) and (ii) for a tree are satisfied and $\mathfrak{B} = (B_0, B_1, \ldots)$ is a tree. The conclusion of the proof 4) $\Rightarrow$ 1) is indirect. We assume that this tree is infinite. Then by the König Graph Theorem 5.4 there exists an infinite branch

$$(\ldots, i_n, i_{n-1}, \ldots, i_1, i_0). \tag{71}$$

For these indices

$$e_{i_n}\varphi_n e_{i_{n-1}}\varphi_{n-1}\cdots e_{i_1}\varphi_{i_1}e_{i_0}(x) \neq 0. \tag{72}$$

Using the factorization $e_i = \iota_i\pi_i$, (72) becomes

$$\iota_{i_n}(\pi_{i_n}\varphi_n\iota_{i_{n-1}})(\pi_{i_{n-1}}\varphi_{n-1}\iota_{i_{n-2}})\cdots(\pi_{i_1}\varphi_1\iota_{i_0})\pi_{i_0}(x) \neq 0.$$

Since $f \in \operatorname{Tot}(S)$, also $\varphi_j = (1-d_j)f(1-d_jf)^{-1} \in \operatorname{Tot}(S)$, and then also

$$\pi_{i_j}\varphi_j\iota_{i_{j-1}} \in \operatorname{Tot}(M_{i_{j-1}}, M_{i_j}).$$

We prove by induction that all the entries in (71) are different. For $n = 0$ there is nothing to prove. We know by (63) and (65) that

$$i_{n-1}, i_{n-2}, \ldots, i_1, i_0 \in K_{n-1}.$$

But

$$i_n \in \operatorname{spt}(\varphi_n e_{i_{n-1}}\cdots e_{i_1}\varphi_1 e_{i_0}(x)),$$

so by (67) $i_n \notin K_{n-1}$.

Now we can apply our lstn assumption to the infinite sequence $f_n = \pi_n\varphi_n\iota_{i_{n-1}}$, $n \in \mathbb{N}$, and the element $\pi_{i_0}(x) \in M_{i_0}$. Accordingly, there for all $n \in \mathbb{N}$ we have $f_nf_{n-1}\cdots f_1(\pi_{i_0}(x)) \neq 0$ contradicting (46). Hence our assumption that $\mathfrak{B}$ was infinite was wrong. Therefore there exists $B_n = \emptyset$ and by (63) $\varphi_n\varphi_{n-1}\cdots\varphi_1(x) = 0$ which was to be shown. With this the Main Theorem 4.1 is proved. □

If we look back, we might observe that there are very interesting relations between the total and other notions. On first sight, this is not obvious. In fact, it requires difficult considerations to prove these connections.

6 Further equivalent conditions

With the Main Theorem 4.1 the list of equivalent conditions is not exhausted. We will add more, some with and some without proof.

The next condition provides information about $1_M + f$ for $f \in \operatorname{Tot}(S)$. We know already that $1_M + f$ is always injective (see (60)), but for $f \notin \operatorname{Rad}(S)$, the map $1_M + f$ is not injective. We will see now that it is "nearly surjective".

Assume again that $M = \bigoplus_{i \in I} M_i$ is an LE–decomposition and set $S := \operatorname{End}(M)$.

Definition 6.1. A submodule $U \subseteq M$ is a **full submodule of** M if and only if

(i) U has an LE–decomposition,

(ii) for the inclusion $\iota : U \to M$ there exists a map $g \in \operatorname{Hom}_R(M, U)$ such that

$$\overline{\iota g} = \overline{1_M}, \quad \overline{g\iota} = \overline{1_U} \quad \text{with}$$

$$\overline{\iota g}, \overline{1_M} \in \overline{S} = S/\operatorname{Tot}(S), \quad \overline{g\iota}, \overline{1_U} \in \operatorname{End}(U)/\operatorname{Tot}(\operatorname{End}(U)).$$

Theorem 6.2. *With the same assumptions on M as before, the following statements hold.*

1) *For any $f \in \operatorname{Tot}(S)$ the module $\operatorname{Im}(1_M + f)$ has an LE–decomposition and $\operatorname{Im}(1_M + f)$ is a full submodule of M that is isomorphic to M.*

2) *If U is a full submodule of M, then there exists $f \in \operatorname{Tot}(S)$ such that*

$$\operatorname{Im}(1_m + f) \subseteq U. \tag{73}$$

3) $\operatorname{Rad}(S) = \operatorname{Tot}(S)$ *if and only if M is the only full submodule of M.*

Proof. 1) In the proof of 4) $\Rightarrow$ 1) in Section 5, we saw that $1_M + f$ is a monomorphism. Hence $\operatorname{Im}(1_M + f)$ is isomorphic to M and has the LE–decomposition

$$(1_M + f)(M) = \bigoplus_{i \in I} (1_M + f)(M_i).$$

Set $U := (1_M + f)(M)$ and

$$g : M \ni x \mapsto (1_M + f)(x) \in U.$$

Then it follows for any $x \in M$ (where $\iota : U \to M$ is the inclusion) that

$$(\iota g - 1_M)(x) = x + f(x) - x = f(x)$$

and therefore

$$\iota g - 1_M = f \tag{74}$$

hence $\overline{\iota g} = \overline{1_M}$ (since $\overline{f} = 0$). As $1_M + f$ is a monomorphism, g is an isomorphism. Then by (74)

$$gfg^{-1} = g(\iota g - 1_M)g^{-1} = g\iota - gg^{-1} = g\iota - 1_U,$$

and, since $f \in \operatorname{Tot}(S)$, this implies that $\overline{g\iota} = \overline{1_U}$.

2) By assumption 6.1.(ii) holds, i.e.,

$$\overline{\iota g} = \overline{1_M}$$

and this implies that there is $f \in \mathrm{Tot}(S)$ such that

$$\iota g = 1_M + f.$$

Therefore,

$$\mathrm{Im}(1_M + f) = \mathrm{Im}(\iota g) = \mathrm{Im}(g) \subseteq U.$$

3) Assume first that $\mathrm{Rad}(S) = \mathrm{Tot}(S)$ and let U be a full submodule of M. Then by 2) there exists $f \in \mathrm{Tot}(S)$ with (73). Since $\mathrm{Rad}(S) = \mathrm{Tot}(S)$, the map $1_M + f$ is an automorphism, hence $M = \mathrm{Im}(1_M + f) \subseteq U$ and $U = M$.

Assume, conversely, that M is the only full submodule of M. Since by 1) for any $f \in \mathrm{Tot}(S)$, $\mathrm{Im}(1_M + f)$ is a full submodule of M, by assumption, $\mathrm{Im}(1_M + f) = M$. Hence $1_M + f$ is also surjective and so an automorphism. But this implies that $f \in \mathrm{Rad}(S)$, so $\mathrm{Tot}(S) \subseteq \mathrm{Rad}(S)$, so $\mathrm{Tot}(S) = \mathrm{Rad}(S)$. □

We state without proof some further conditions equivalent to $\mathrm{Rad}(S) = \mathrm{Tot}(S)$. It is not difficult to verify these using the main theorem and its proof but we demonstrated sufficiently how the total works in these contexts.

Remark. Let M be a module with LE–decompositions. Then the following statements are equivalent.

1) $\mathrm{Rad}(S) = \mathrm{Tot}(S)$.

2) S is semiregular which means

   i) $S/\mathrm{Rad}(S)$ is regular and

   ii) idempotents in $S/\mathrm{Rad}(S)$ lift to S.

3) Every independent family that finitely generates direct summands and consists of submodules of M with local endomorphism rings, is itself a direct summand of M.

Recall that an independent family of submodules "finitely generates direct summands" if every finite subfamily generates a direct summand.

As mentioned before we give no proofs but we point out one interesting item. For the proof of 1) $\Rightarrow$ 2), the construction in Section V.3 had to be generalized. If M and N are two modules with LE–decompositions, then $\mathrm{Hom}_R(M,N)/\mathrm{Tot}(M,N)$ can be considered a product of Hom's between vector spaces, similar to Section V.3 in the case $M = N$.

With this remark we conclude our presentation of some fundamental properties of partially invertible elements and the total.

Chapter V

The total in torsion–free Abelian groups

1 Background

In this chapter "group" means torsion–free Abelian group. The "bible" of Abelian group theory is the two volume work of Fuchs [11]. Other important sources for torsion–free group are the books by Arnold [2], [1] and all that is needed here can be found (with proof) in [25].

Abelian groups are $\mathbb{Z}$–modules, and any group homomorphism is a $\mathbb{Z}$–module homomorphism. It is customary to write $\mathrm{Hom}(A, B)$ instead of $\mathrm{Hom}_Z(A, B)$ in works on Abelian groups. An Abelian group G is **torsion–free** if for $0 \neq x \in G$ and $m \in \mathbb{Z}$, we have $mx = 0$ if and only if $m = 0$. Every torsion–free Abelian group G is embedded in a $\mathbb{Q}$–vector space as an additive subgroup. Formally such an embedding can be achieved via the tensor product: $G \cong \mathbb{Z} \otimes_{\mathbb{Z}} G \rightarrowtail \mathbb{Q} \otimes_{\mathbb{Z}} G$. We therefore may and do think of torsion–free Abelian groups as additive subgroups of $\mathbb{Q}$–vector spaces. In particular, if G is a torsion–free group and $x \in G$, then there is a well–defined scalar product rx, $r \in \mathbb{Q}$, although rx need not belong to G but only to the vector space in which G is embedded. In any embedding in a $\mathbb{Q}$–vector space, $\mathbb{Q}G$ denotes the subspace spanned by G, and it is easy to see that $\mathbb{Q}G = \{rx \mid r \in \mathbb{Q}, x \in G\} = \{\frac{1}{n}x \mid n \in \mathbb{N}, x \in G\}$. The **rank** of a torsion–free Abelian group G, $\mathrm{rk}(G)$, is the dimension of the $\mathbb{Q}$–space spanned by G: $\mathrm{rk}(G) = \dim(\mathbb{Q}G)$.

We note that

$$\mathrm{rk}(G \oplus H) = \mathrm{rk}(G) + \mathrm{rk}(H). \tag{1}$$

Proof. It is clear that $\mathbb{Q}(G \oplus H) = \mathbb{Q}G + \mathbb{Q}H$. We need to show that with $G \cap H = 0$, also $\mathbb{Q}G \cap \mathbb{Q}H = 0$. Let $x \in \mathbb{Q}G \cap \mathbb{Q}H$. Then $x = \frac{1}{n}g = \frac{1}{n'}h$ for rational numbers $\frac{1}{n}, \frac{1}{n'}$, and elements $g \in G$, $h \in H$. It follows that $n'g = nh \in G \cap H = 0$. As our groups are torsion–free, it follows that $g = h = 0$, so $x = 0$. □

The **rational groups** are the additive subgroups of $\mathbb{Q}$ that contain $\mathbb{Z}$. The requirement that a rational group contain $\mathbb{Z}$ is convenient. We will denote the rational groups by $A, , B, \ldots$. It is easy to see that every additive subgroup of $\mathbb{Q}$ is isomorphic to a rational group. (If $0 \neq \frac{m}{n} \in \tau$, then $\frac{n}{m}\tau \cong \tau$ and $1 \in \frac{n}{m}\tau$) Examples of rational groups are the subrings $\mathbb{Z}[p^{-1}] = \{\frac{m}{p^n} \mid m \in \mathbb{Z}, n \in \mathbb{N}_0\}$ of $\mathbb{Q}$. Every rational group A is directly indecomposable because of (1) and $\mathbb{Q}A = \mathbb{Q}$, so $\mathrm{rk}(A) = 1$. However, as we will see soon, rational groups need not have local endomorphism rings. We begin by studying homomorphisms between rational groups.

By $(m/n)^\bullet$ we mean multiplication by the rational number m/n.

Lemma 1.1. *Let A, B be rational groups.*

1) *If $f \in \mathrm{Hom}(A, B)$, then f extends uniquely to a linear mapping of $\mathbb{Q}$ and hence is multiplication by a rational number, namely $f(1)$.*
2) *Non–zero homomorphisms between rational groups are injective. Non–zero homomorphisms $A \to B$ exist if and only if $mA \subseteq B$ for some non–zero integer m. Furthermore,*

$$\mathrm{Hom}(A, B) = \{(m/n)^\bullet \mid mA \subseteq nB\}.$$

3) $\mathrm{End}(A) = \{(m/n)^\bullet : nA = A\}$.

Proof. 1) $\{1\}$ is a basis of $\mathbb{Q}$ as a $\mathbb{Q}$–vector space, so by linear algebra there is a unique linear mapping $g : \mathbb{Q} \to \mathbb{Q}$ such that $g(1) = f(1)$. We claim that g restricts to f on A. Let $0 \neq x \in A$. Then $x = m/n$ for some integers m, n. So $nx = m \in A$ and

$$g(x) = g\left(\frac{m}{n}\right) = \frac{m}{n}g(1) = \frac{m}{n}f(1) = \frac{1}{n}f(m) = \frac{1}{n}f(nx) = f(x),$$

hence $f(x) = xf(1)$.

2) Clearly, multiplication by a non–zero rational number is an injective map. If $0 \neq (m/n)^\bullet \in \mathrm{Hom}(A, B)$, then $(m/n)A \subseteq B$ and hence $mA \subseteq nB \subseteq B$. Conversely, if $mA \subseteq nB$, then multiplication by m/n maps A into B.

3) Let $(m/n)^\bullet \in \mathrm{End}(A)$ and assume without loss of generality that $\gcd(m, n) = 1$. Then we have integers u, v such that $1 = um + vn$. We conclude that $\frac{1}{n}A \subseteq u\frac{m}{n}A + vA \subseteq A$. Thus $A \subseteq nA \subseteq A$, so $nA = A$. The converse is immediate. □

In the following we will drop the bullet from the multiplication $(m/n)^\bullet$ and make the identification $(m/n)^\bullet = m/n$. It will be clear from the context whether m/n is a fraction or multiplication by m/n.

Example 1.2. Let p be a prime number. The additive group of the ring $\mathbb{Z}[p^{-1}]$ is directly indecomposable but $S := \mathrm{End}(\mathbb{Z}[p^{-1}]) \cong \mathbb{Z}[p^{-1}]$ is not a local ring. The group of units of S, i.e., $\mathrm{Aut}(\mathbb{Z}[p^{-1}])$, is the multiplicative group $\{\pm p^n \mid n \in \mathbb{Z}\}$. Moreover, $\mathrm{Rad}(\mathbb{Z}[p^{-1}]) = 0$ and $\mathrm{Tot}(\mathbb{Z}[p^{-1}]) = \mathbb{Z}[p^{-1}] \setminus \mathrm{Aut}(\mathbb{Z}[p^{-1}]) = \bigcup_{q \neq p} \mathbb{Z}[p^{-1}]q$.

Proof. Recall that $\mathbb{Z}[p^{-1}] = \{m/p^n \mid m, n \in \mathbb{Z}\}$. An endomorphism of $\mathbb{Z}[p^{-1}]$ is multiplication by a rational number r that must be of the form m/p^n because $r\mathbb{Z}[p^{-1}] \subseteq \mathbb{Z}[p^{-1}]$. Automorphisms are endomorphisms (= rationals) whose inverse is again an endomorphism. This leaves exactly the (positive or negative) powers of p. The maximal ideals of $\mathbb{Z}[p^{-1}]$ are exactly the subsets $q\mathbb{Z}[p^{-1}]$ where q is a prime different from p. The intersection of all these maximal ideals is $\{0\}$, so $\mathrm{Rad}(S) = \{0\}$. The identity 1 is the only non–zero idempotent in S, hence the pi elements are exactly the invertible elements and $\mathrm{Tot}(\mathbb{Z}[p^{-1}]) = \mathbb{Z}[p^{-1}] \setminus \mathrm{Aut}(\mathbb{Z}[p^{-1}])$. It is easy to verify that $\mathbb{Z}[p^{-1}] \setminus \mathrm{Aut}(\mathbb{Z}[p^{-1}]) = \bigcup_{q \neq p} \mathbb{Z}[p^{-1}]q$: Suppose first that $mp^a \notin \{\pm p^n \mid n \in \mathbb{Z}\}$ where $a \in \mathbb{Z}$ and $m \in \mathbb{Z}$, $\gcd(m, p) = 1$. Then m must have a prime factor $q \neq p$ and $mp^a \in q\mathbb{Z}[p^{-1}]$. Conversely, if $x \in q\mathbb{Z}[p^{-1}]$, then surely $x \notin \{\pm p^n \mid n \in \mathbb{Z}\}$. □

We insert here a fact that is very helpful and will be used in the following without explicit reference.

Lemma 1.3. *Let $f \in \mathrm{Hom}(G, H)$. Then ϕ has a unique extension $g : \mathbb{Q}G \to \mathbb{Q}H$ that is a linear mapping. Hence any homomorphism $f : G \to H$ may be considered to be the restriction of a linear mapping $g : \mathbb{Q}G \to \mathbb{Q}H$.*

Proof. Let $f \in \mathrm{Hom}(G, H)$ be given. We define $g : \mathbb{Q}G \to \mathbb{Q}H$ as follows. For $x = \frac{1}{n}y \in \mathbb{Q}G$, $y \in G$, let $g(x) = \frac{1}{n}\phi(y) \in \mathbb{Q}H$. It is an easy exercise to check that g is well–defined (independent of the choice of n and y) and linear. □

We will use the following well–known result from number theory.

Lemma 1.4 (Partial Fraction Decomposition). *Let n be a positive integer and let $n = \prod_p p^{n_p}$ be its prime factorization. Then there exist integers u_p such that*

$$\frac{1}{n} = \sum_p \frac{u_p}{p^{n_p}}.$$

Rational groups can be obtained by choosing a set of generators. For example, $\mathbb{Z} = \langle 1 \rangle$ and $\mathbb{Q} = \left\langle \frac{1}{p^n} : p \in \mathbb{P}, n \in \mathbb{N}_0 \right\rangle$. The following lemma produces simplified generating sets and some applications.

Lemma 1.5. *Let A and B be rational groups.*

1) *If $\frac{m}{n} \in A$ with $\gcd(m, n) = 1$, then $\frac{1}{n} \in A$ and A is generated by its elements of the form $\frac{1}{p^k}$ where $p \in \mathbb{P}$.*

2) *If $\frac{1}{m}, \frac{1}{n} \in A$, then $\frac{1}{\mathrm{lcm}(m,n)} \in A$.*

3) *If $1/p_i^{m_i} \in A$ for $1 \leq i \leq I$, then $1/\prod_{1 \leq i \leq I} p_i^{m_i} \in A$.*

4) *$A \subseteq B$ if and only if whenever $1/p^k \in A$, then $1/p^k \in B$.*

5) *If $B = mA$, and $\gcd(m, n) = 1$, then $1/n \in B$ if and only if $1/n \in A$.*

Proof. 1) Assume that $\frac{m}{n} \in A$ with $\gcd(m,n) = 1$. Choose integers u, v such that $um + vn = 1$. Then $\frac{1}{n} = u\frac{m}{n} + v \in A$. Hence A is generated by its fractions $\frac{1}{n}$. (Here our convention that a rational group contains $\mathbb{Z}$ is used) Let $\frac{1}{n}$ be such a generator and consider its prime factorization $n = \prod p^{n_p}$. By Lemma 1.4

$$\frac{1}{n} = \sum_p \frac{u_p}{p^{n_p}} \in A.$$

This shows that the special fractions $1/p^{n_p} \in A$ generate A.

2) Write $\gcd(m,n) = um + vn$. Then

$$\frac{1}{\operatorname{lcm}(m,n)} = \frac{\gcd(m,n)}{mn} = u\frac{1}{n} + v\frac{1}{m} \in A.$$

3) Apply 2) repeatedly.

4) By 1) A is generated by negative prime powers.

5) Suppose that $B = mA$ and $\gcd(m,n) = 1$. Let $1/n \in B$. Then $1/n \in mA \subseteq A$. Conversely, let $1/n \in A$ and write $1 = um+vn$. Then $\frac{1}{n} = u\left(m\frac{1}{n}\right) + v \in B$. □

The following lemma will enter our discussion at a crucial point.

Lemma 1.6. *Any bounded quotient of a rational group is cyclic.*

Proof. One checks that $\mathbb{Q}/\mathbb{Z} \cong \bigoplus_{p\in\mathbb{P}} \mathbb{Z}(p^\infty)$ where $\mathbb{Z}(p^\infty)$ is the Prüfer group defined in Chapter II, Example 6.3. We will use below that every epimorphic image of $\mathbb{Q}/\mathbb{Z}$ is of the form $\bigoplus_{p\in\mathbb{P}'} \mathbb{Z}(p^\infty)$ where $\mathbb{P}'$ is a certain subset of $\mathbb{P}$. Consequently, every subgroup of $\mathbb{Q}/\mathbb{Z}$ and any of its epimorphic images is of the form $\bigoplus_{p\in\mathbb{P}'} U_p$ where U_p is a subgroup of $\mathbb{Z}(p^\infty)$, so is cyclic or the whole group $\mathbb{Z}(p^\infty)$. Any bounded subgroup of $\mathbb{Q}/\mathbb{Z}$ or any of its epimorphic images is cyclic.

Let e be a positive integer. Then A/eA is cyclic. In fact,

$$\frac{\mathbb{Q}}{\mathbb{Z}} \twoheadrightarrow \frac{\mathbb{Q}}{A} \cong \frac{e\mathbb{Q}}{eA} = \frac{\mathbb{Q}}{eA} \geq \frac{A}{eA},$$

hence A/eA is cyclic. Suppose that A/K is bounded by e. Then $eA \subset K$ and there is a natural epimorphism $A/eA \twoheadrightarrow A/K$. Since A/eA is cyclic, so is A/K. □

We now turn to the problem of classification. The first lemma is crucial and powerful.

Lemma 1.7. *Let A, B be rational groups. Then the following are equivalent.*

1) $A \cong B$,

2) *there exist non–zero homomorphisms $A \to B$ and $B \to A$,*

3) *there exist integers m, n such that $mA \subset B$ and $nB \subset A$.*

Proof. 1) $\Rightarrow$ 2): Any given isomorphism and its inverse will do.

2) $\Rightarrow$ 3): If $A \to B$ is a non–zero homomorphism, then it is multiplication by some reduced fraction $\frac{m}{n}$ and so $\frac{m}{n}A \subset B$. This says that $mA \subset nB \subset B$. The same argument applies with the roles of A and B reversed.

3) $\Rightarrow$ 1): Since $mA \subset B$ and $nB \subset A$, it follows that $mnA \leq nB \leq A$ and $nB \cong B$. Hence we assume without loss of generality that $n = 1$, i.e., $mA \leq B \leq A$. The group A/mA is cyclic by Lemma 1.6. The subgroups of the cyclic group A/mA are of the form $k(A/mA)$ where k is a factor of $|A/mA|$ which in turn is a factor of m. So $k(A/mA) = kA/mA$ and $B/mA = kA/mA$ for some k. Therefore $B = kA \cong A$. □

A group G of rank 1 is an additive subgroup of $\mathbb{Q}G \cong \mathbb{Q}$, and hence is isomorphic to a rational group. A **type** is an isomorphism class of rank–one groups. We denote the types by $\sigma, \tau, \rho, \ldots$. Rational groups have rank 1 and every type is represented by a rational group (use that $\mathbb{Q}$ is injective and every non–zero cyclic subgroup of a rank–one group is large). The set of all types $\mathbb{T}$ (there are $2^{\aleph_0}$ of them) can be partially ordered as follows.

Definition 1.8. Let $\sigma, \tau \in \mathbb{T}$. Choose rank-one groups A and B with $\mathrm{tp}(A) = \sigma$ and $\mathrm{tp}(B) = \tau$. Set $\sigma \leq \tau$ if and only if $\mathrm{Hom}(A, B) \neq 0$.

Lemma 1.9. *The relation $\leq$ is a well–defined partial order on the set of types $\mathbb{T}$.*

Proof. It is immediate that the relation does not depend on the choice of the representatives A and B, and it is also clear that the relation is reflexive and transitive. The main problem is to show that $\sigma = \tau$ follows from $\sigma \leq \tau$ and $\tau \leq \sigma$. But this is settled by Lemma 1.7. □

We use $\sigma < \tau$, $\sigma \leq \tau$, $\sigma > \tau$ and $\sigma \geq \tau$ as usual. The smallest type is the class of $\mathbb{Z}$ and the largest type is the class of $\mathbb{Q}$. With its partial order $\mathbb{T}$ is a lattice: given $\sigma, \tau \in \mathbb{T}$ choose A, B such that $\mathrm{tp}(A) = \sigma$ and $\mathrm{tp}(B) = \tau$. Then $\sigma \wedge \tau = \mathrm{tp}(A \cap B)$ and $\sigma \vee \tau = \mathrm{tp}(A + B)$.

A **completely decomposable group** is a torsion–free Abelian group that is a direct sum of groups of rank 1.

Decompositions of completely decomposable groups with indecomposable summands are not LE–decompositions, yet they are unique up to isomorphism. To show this uniqueness of decompositions we need to introduce the so–called τ–socle of a torsion–free Abelian group. It requires no extra effort to treat the subject in greater generality.

Definition 1.10. Let A be a group. The A–**socle** of a group G is defined as

$$\mathrm{Soc}_A(G) \quad = \quad \sum \{f(A) \mid f \in \mathrm{Hom}(A, G)\}.$$

For a given type τ, choose a rational group A of type τ. The τ–**socle** of G is $G(\tau) := \mathrm{Soc}_A(G)$.

The A–socle clearly depends only on the isomorphism class of A, so the τ–socle $G(\tau)$ is well–defined.

The following lemma shows that A–socles are justifiably called socles and are so–called functorial subgroups, a term introduced by B. Charles [9]. A **functorial subgroup** in the category of torsion–free Abelian groups is a functor S that assigns to every torsion–free Abelian group G a subgroup $\mathrm{S}(G)$ such that $f(\mathrm{S}(G)) \subseteq \mathrm{S}(H)$ for every $f \in \mathrm{Hom}(G,H)$. In particular, functorial subgroups are fully invariant, i.e., for all $f \in \mathrm{End}(G)$, we have that $f(\mathrm{S}(G)) \subseteq \mathrm{S}(G)$.

Lemma 1.11. *Let A, G, H be torsion–free groups and $f \in \mathrm{Hom}(G,H)$. Then the following statements hold.*

1) $\mathrm{Soc}_A(\mathrm{Soc}_A(G)) = \mathrm{Soc}_A(G)$.

2) $f(\mathrm{Soc}_A(G)) \subseteq \mathrm{Soc}_A(H)$. *Thus socles are functorial subgroups.*

Proof. We will set $\mathrm{Hom}(A,G)(A) := \sum\{f(A) : f \in \mathrm{Hom}(A,G)\}$.

1) The inclusion $\mathrm{Soc}_A(G) \rightarrowtail G$ induces an isomorphism

$$\mathrm{Hom}(A, \mathrm{Soc}_A(G)) \cong \mathrm{Hom}(A,G),$$

and hence

$$\mathrm{Soc}_A(\mathrm{Soc}_A(G)) = \mathrm{Hom}(A, \mathrm{Soc}_A(G))(A) = \mathrm{Hom}(A,G)(A) = \mathrm{Soc}_A(G).$$

2) $f(\mathrm{Soc}_A(G)) = f(\mathrm{Hom}(A,G)(A)) \subseteq \mathrm{Hom}(A,H)(A) = \mathrm{Soc}_A(H)$. □

The following observations will be applied to A–socles but they are conveniently proved for functorial subgroups in general.

Lemma 1.12. *Let* S *be a functorial subgroup on the category of torsion–free Abelian groups.*

1) *If $H \subseteq G$, then $\mathrm{S}(H) \subseteq H \cap \mathrm{S}(G)$.*

2) *Suppose that H is a subgroup of G such that there exists an endomorphism of G that is the identity on H and whose image is contained in H. Then $\mathrm{S}(H) = H \cap \mathrm{S}(G)$. This is true in particular when H is a direct summand of G.*

3) *Let G, H be torsion–free Abelian groups. Then*

$$\mathrm{S}(G \oplus H) = \mathrm{S}(G) \oplus \mathrm{S}(H).$$

Proof. 1) Let $f : H \to G$ be the inclusion. Then by definition of functorial subgroup, $\mathrm{S}(H) \subseteq H \cap \mathrm{S}(G)$.

2) Let $f \in \mathrm{End}(G)$ such that $f\restriction_H = 1$ and $f(G) \subseteq H$. Then

$$\mathrm{S}(H) \subseteq H \cap \mathrm{S}(G) = f(H \cap \mathrm{S}(G)) \subseteq f(\mathrm{S}(G)) \subseteq \mathrm{S}(H)$$

where $H \cap \mathrm{S}(G) = f(H \cap \mathrm{S}(G))$ since f is the identity on H.

3) Set $K = G \oplus H$. Since $\mathrm{S}(K)$ is fully invariant in K, it follows by Lemma 0.3 that

$$\mathrm{S}(K) = G \cap \mathrm{S}(K) \oplus H \cap \mathrm{S}(K),$$

and by 2), applied to the projections belonging to this decomposition, we obtain that $G \cap \mathrm{S}(K) = \mathrm{S}(G)$ and $H \cap \mathrm{S}(K) = \mathrm{S}(H)$. □

As specific examples we mention that $G(\mathbb{Z}) = G$ and $G(\mathbb{Q})$ is the unique maximal divisible (= injective) subgroup of G.

Let x be an element of a group $G \subseteq \mathbb{Q}G$. If $r \in \mathbb{Q}$, then $rx \in \mathbb{Q}G$ but rx may or may not be in G. We define $\mathbb{Q}_x^G = \{r \in \mathbb{Q} \mid rx \in G\}$. Then $\mathbb{Q}_x^G$ is a rational group, called the **coefficient group of** x. The **type of an element** $x \in G$, denoted $\mathrm{tp}^G(x)$ is the isomorphism class of $\mathbb{Q}_x^G$, i.e., $\mathrm{tp}^G(x) = \mathrm{tp}(\mathbb{Q}_x^G)$.

Furthermore, we need the notion of "purity" that is due to Heinz Prüfer, one of the pioneers of Abelian group theory. In the case of torsion–free groups this is a simple matter: A subgroup H of the torsion–free group G is **pure** in G if and only if G/H is again torsion–free. Equivalently, whenever $0 \neq ng \in H$ for some integer n and some element $g \in G$, then $g \in H$. It is also true that H is pure in G, if and only if for all $x \in H$, there is an equality of coefficient groups $\mathbb{Q}_x^H = \mathbb{Q}_x^G$. It is not difficult to show that any subgroup H of a torsion–free group G is contained in a unique smallest pure subgroup of G, called the **pure hull** of H and denoted by H_*^G. In fact, H_*^G/H is equal to the maximal torsion subgroup of G/H. If $x \in G$, then the pure hull of $\mathbb{Z}x$ in G is $\mathbb{Q}_x^G x := \{rx \mid r \in \mathbb{Q}_x^G\}$.

We will need some of the properties of coefficient groups.

Lemma 1.13. *Let G be a torsion–free group and $x, y \in G$. Then the following hold.*

1) $\mathbb{Q}_{x+y}^G \supset \mathbb{Q}_x^G \cap \mathbb{Q}_y^G$.
2) *For a non–zero integer m,* $\mathbb{Q}_{mx}^G = \frac{1}{m}\mathbb{Q}_x^G$.
3) *Suppose that $G = \bigoplus_i G_i$ and $x = \sum_i x_i \in G$ where $x_i \in G_i$. Then* $\mathbb{Q}_x^G = \bigcap_i \mathbb{Q}_{x_i}^{G_i}$.

Proof. 1) If $r \in \mathbb{Q}_x^G \cap \mathbb{Q}_y^G$, then $rx, ry \in G$, so $r(x+y) = rx + ry \in G$, and $r \in \mathbb{Q}_{x+y}^G$.

2) $r \in \mathbb{Q}_{mx}^G \Leftrightarrow rmx \in G \Leftrightarrow rm \in \mathbb{Q}_x^G \Leftrightarrow r \in \frac{1}{m}\mathbb{Q}_x^G$.

3) Let $x = \sum_i x_i \in G$ with $x_i \in G_i$, and let $r \in \mathbb{Q}$. Then

$$r \in \mathbb{Q}_x^G \Leftrightarrow rx \in G \Leftrightarrow \ (\forall i)\ rx_i \in G_i \Leftrightarrow r \in \bigcap_i \mathbb{Q}_{x_i}^{G_i}.$$ □

The existence of homomorphisms is closely tied in with coefficient groups as the next lemma shows.

Lemma 1.14. *Let G be a torsion–free group and $x \in \mathbb{Q}G$. Then the following hold.*

1) *If $f \in \operatorname{Hom}(G, H)$ and $x \in \mathbb{Q}G$, then $\mathbb{Q}_x^G \subseteq \mathbb{Q}_{f(x)}^H$.*

2) *Let $x \in G$ and $y \in H$. Then there exists a homomorphism $f : \mathbb{Q}_x^G x \to H$ such that $f(x) = y$ if and only if $\mathbb{Q}_x^G \subseteq \mathbb{Q}_y^H$.*

Proof. 1) Recall that f may be considered a linear mapping $\mathbb{Q}G \to \mathbb{Q}H$ (Lemma 1.3). Let $r \in \mathbb{Q}_x^G$. Then $r(f(x)) = f(rx) \in H$, and $r \in \mathbb{Q}_{f(x)}^H$.

2) The necessity follows from 1). For the sufficiency define $f : \mathbb{Q}_x^G x \to H :$ $f(rx) = ry$ for $r \in \mathbb{Q}_x^G$. Since $\mathbb{Q}_x^G \subseteq \mathbb{Q}_y^G$ by hypothesis, it follows that $ry \in H$ and f is a well–defined homomorphism into H. □

The basic properties of types of elements are collected in the following lemma.

Lemma 1.15. *Let G be a torsion–free group and $x, y \in G$.*

1) *If $x \in H$ and H is pure in G, then $\operatorname{tp}^H(x) = \operatorname{tp}^G(x)$.*

2) *If n is a non–zero integer, then $\operatorname{tp}^G(nx) = \operatorname{tp}^G(x)$.*

3) $\operatorname{tp}^G(x + y) \geq \operatorname{tp}^G(x) \wedge \operatorname{tp}^G(y)$.

4) *If $G = \bigoplus_i G_i$, and $x = \sum_i x_i$, $x_i \in G_i$, then $\operatorname{tp}^G(x) = \bigwedge_i \operatorname{tp}^G(x_i)$.*

5) *If $f : G \to H$ is a homomorphism and $x \in G$, then $\operatorname{tp}^H(f(x)) \geq \operatorname{tp}^G(x)$.*

6) *Let A be a rational group and a a non–zero element of A. Then $\operatorname{tp}^A(a) =$ $\operatorname{tp}(A)$.*

Proof. 1) By definition of purity, $\mathbb{Q}_x^H = \mathbb{Q}_x^G$.

2) By Lemma 1.13.2), $n\mathbb{Q}_{nx}^G = \mathbb{Q}_x^G$, hence $\mathbb{Q}_{nx}^G$ and $\mathbb{Q}_x^G$ are isomorphic rational groups and therefore have equal types.

3) Lemma 1.13.1).

4) Lemma 1.14.2).

5) Lemma 1.13.3).

6) $A = \mathbb{Q}_a^A a \cong \mathbb{Q}_a^A$. So $\operatorname{tp}^A(a) = \operatorname{tp}(A)$. □

The following result is fundamental.

Lemma 1.16. *Let $A \subseteq B$ be rational groups. Then $\operatorname{Soc}_A(B) = B$.*

Proof. By Lemma 1.5.4) it is enough to show that $1/p^k \in \operatorname{Soc}_A(B)$ whenever $1/p^k \in B$. Let $1/p^k \in B$. We wish to find an element a of A and a mapping $f \in \operatorname{Hom}(A, B)$ such that $f(a) = 1/p^k$ and will use Lemma 1.14.2) to do so. In the case that A contains all fractions $1/p^m$, the choice $a = 1 \in A$ will do. In this case $\mathbb{Q}_{1/p^k}^B$ also contains all fractions $1/p^m$. Furthermore,

$$p^k\mathbb{Q}_1^A \overset{\text{evident}}{\subseteq} p^k\mathbb{Q}_1^B \overset{1.13.2)}{=} \mathbb{Q}_{1/p^k}^B.$$

Therefore, for primes q different from p we have $1/q^m \in \mathbb{Q}_a^A \overset{1.5.5)}{\Leftrightarrow} 1/q^m \in p^k\mathbb{Q}_a^A$, so if in fact $1/q^m \in \mathbb{Q}_1^A$, then $1/q^m \in \mathbb{Q}_{1/p^k}^B$. This shows that the desired map $f : A = \mathbb{Q}_1^A 1 \to B$ with $f(1) = 1/p^k$ does exist. In the case that A does not contain all fractions $1/p^m$ we choose $a = 1/p^n \in A$ where n is maximal with this property. Then $\mathbb{Q}_{1/p^n}^A$ contains no proper fraction $1/p^m$ and that $1/q^m \in \mathbb{Q}_{1/p^n}^A$ implies that $1/q^m \in \mathbb{Q}_{1/p^k}^B$ follows as above from Lemma 1.5.5) and

$$p^k\mathbb{Q}_{1/p^n}^A \overset{\text{evident}}{\subseteq} p^k\mathbb{Q}_{1/p^n}^B \overset{1.13.2)}{=} p^n\mathbb{Q}_{1/p^k}^B \subseteq \mathbb{Q}_{1/p^k}^B.$$

Again we have a map $f : A = \mathbb{Q}_{1/p^n}^A(1/p^n) \to B$ with $f(1/p^n) = 1/p^k$. □

We now obtain a powerful description of socles.

Corollary 1.17. *Let G be a torsion–free Abelian group and τ, σ types.*

1) $G(\tau) = \{x \in G \mid \mathrm{tp}^G(x) \geq \tau\}$.
2) $G(\tau)$ *is pure in* G.
3) *If* $\sigma \geq \tau$, *then* $G(\sigma) \subseteq G(\tau)$.

Proof. 1) Choose a rational group A such that $\mathrm{tp}(A) = \tau$. Suppose that $x \in G(\tau)$. Then there exist maps $f_i \in \mathrm{Hom}(A, G)$ and elements $a_i \in A$ such that $x = \sum_i f_i(a_i)$. Hence

$$\mathrm{tp}^G(x) \overset{1.15.3)}{\geq} \bigwedge_i \mathrm{tp}^G(f_i(a_i) \overset{1.15.5)}{\geq} \bigwedge_i \mathrm{tp}^G(a_i) \overset{1.15.6)}{\geq} \tau.$$

2) Part 1) and Lemma 1.15.2).

3) Immediate consequence of 1). □

We are now ready to introduce the announced additional type subgroups.

Definition 1.18. Let G be a torsion–free Abelian group and let σ, ρ be types. Let

$$G^\sharp(\sigma) = \left(\sum_{\rho > \sigma} G(\rho)\right)_*^G.$$

For easy reference we list some properties of these type subgroups in a lemma. They either follow directly from the definition or from earlier lemmas on functorial subgroups.

Lemma 1.19. *Let G be a torsion–free group.*

1) $G(\sigma) \supseteq G^\sharp(\sigma)$.
2) *If* $f \in \mathrm{Hom}(G, H)$, *then* $f(G(\sigma)) \subseteq H(\sigma)$, *and* $f(G^\sharp(\sigma)) \subseteq H^\sharp(\sigma)$, *i.e., all type subgroups are functorial subgroups.*

3) *If* $f : G \to H$ *is an isomorphism, then* $f \restriction_{G(\sigma)} : G(\sigma) \to H(\sigma)$ *and* $f \restriction_{G^\sharp(\sigma)}$: $G^\sharp(\sigma) \to H^\sharp(\sigma)$ *are isomorphisms.*

4) *If* $G = G_1 \oplus G_2$, *then*

$$G(\tau) = G_1(\tau) \oplus G_2(\tau), \quad G^\sharp(\tau) = G_1^\sharp(\tau) \oplus G_2^\sharp(\tau).$$

5) *If* $eG \subset H \subset G$ *for some positive integer* e, *then* $H(\tau) = H \cap G(\tau)$ *and* $G(\tau) = H(\tau)_*^G$.

A particular decomposition of completely decomposable groups will be introduced next and used routinely for the rest of this chapter. At the same time the type subgroups are described and invariants of the particular decomposition are found.

Definition 1.20. A torsion–free group G is called τ–**homogeneous** if for every non–zero element $x \in G$, it is true that $\mathrm{tp}^G(x) = \tau$.

Lemma 1.21. *Let* A *be a completely decomposable group.*

1) *There is a decomposition* $A = \bigoplus_{\rho\in\mathbb{T}} A_\rho$ *where each summand* A_τ *is* τ–*homogeneous completely decomposable.*

2) *If* $A = \bigoplus_{\rho\in\mathbb{T}} A_\rho$ *is a decomposition such that each summand* A_τ *is* τ–*homogeneous completely decomposable, then*

$$A(\tau) = \bigoplus_{\rho\geq\tau} A_\rho, \quad \text{and} \quad A^\sharp(\tau) = \bigoplus_{\rho>\tau} A_\rho.$$

Consequently, $A(\tau)/A^\sharp(\tau) \cong A_\tau$.

Proof. 1) Collect summands of the same type.

2) Use Corollary 1.17.1) and Lemma 1.15.4). □

Completely decomposable groups were classified by Baer ([3]). The classification is an immediate consequence of Lemma 1.21.

Theorem 1.22. (Reinhold Baer 1937) *Let* A *be a completely decomposable group. In any direct decomposition of* A *into rank–one summands, the number of summands having type* τ *is given by*

$$\mathrm{rk}\left(A(\tau)/A^\sharp(\tau)\right).$$

Hence these cardinals form a complete set of isomorphism invariants for A.

We introduce terminology in order to simplify the discourse.

Definition 1.23. Let A be completely decomposable and $A = \bigoplus_{\rho\in\mathbb{T}} A_\rho$ be a direct decomposition into τ–homogeneous completely decomposable summands A_τ. We call the A_τ τ–**homogeneous components** of A and $A = \bigoplus_{\rho\in\mathbb{T}} A_\rho$ a **homogeneous decomposition** of A.

Only the types for which $A(\tau)/A^\sharp(\tau)$ is non–zero are of significance in a homogeneous decomposition of a completely decomposable group, which suggests the following definitions.

Definition 1.24. Let G be any torsion–free group. The type τ is a **critical type** of G if

$$G(\tau)/G^\sharp(\tau) \neq 0.$$

Let $\mathrm{T}_{\mathrm{cr}}(G)$ denote the set of all critical types of G.

We conclude this section with a number of classical results that are important tools.

Proposition 1.25. (Baer [3], Kolettis [23, Theorem 7.]) *Suppose that A is τ–homogeneous completely decomposable of arbitrary rank. If B is a τ–homogeneous subgroup of A (in particular if B is pure in A), then B is completely decomposable.*

Proof. [11, Theorem 86.6] or [25, Proposition 2.4.14]. □

The next theorem is the Baer–Kulikov–Kaplansky Theorem.

Theorem 1.26. *Every direct summand of a completely decomposable group is again completely decomposable.*

Proof. [11, Theorem 86.7]) for the general version or [25, Theorem 2.4.15] for the finite rank version. □

The following is a frequently used proposition.

Proposition 1.27. *A pure subgroup of a homogeneous completely decomposable group of finite rank is a direct summand and again completely decomposable.*

Proof. [11, Lemma 86.8]) or [25, Proposition 2.4.17]. □

Another important result in the theory of torsion–free groups of finite rank is the following proposition. The symbol $[G : H]$ denotes the **index** of the subgroup H of G in G, in other words, $[G : H]$ is the cardinality of the quotient group G/H.

Proposition 1.28. *Let G and H be torsion–free groups of finite rank.*

1) *If $f \in \operatorname{End} G$ is injective, then $[G : f(G)]$ is finite.*
2) *If there is a monomorphism $f : G \to H$ and a monomorphism $g : H \to G$, then the indices $[H : f(G)]$ and $[G : g(H)]$ are finite.*

Proof. [25, Proposition 2.1.3]. □

2 The total of completely decomposable groups

Our first theorem of this section contains a description of the total of a homogeneous completely decomposable group of finite rank. The symbol $\mathbb{P}$ denotes the set of all prime numbers.

Theorem 2.1. *Let A be a τ–homogeneous completely decomposable group of finite rank and $f \in \operatorname{End}(A)$. Then $f \in \operatorname{Tot}(\operatorname{End}(A))$ if and only if there exists a prime p such that $f(A) \subseteq pA \neq A$. Consequently,*

$$\operatorname{Tot}(\operatorname{End}(A)) = \bigcup \{q \operatorname{End}(A) \mid q \in \mathbb{P}, qA \neq A\}.$$

Proof. We will show that any $f \in \operatorname{End}(A)$ is pi if $f(A) \not\subseteq pA$ whenever $pA \neq A$.

Let $f \in \operatorname{End}(A)$. The kernel $\operatorname{Ker}(f)$ is a pure subgroup of A since $A/\operatorname{Ker}(f) \cong f(A) \subseteq A$ is torsion–free. Hence $A = \operatorname{Ker}(f) \oplus K$ (Proposition 1.27) and $K \ni x \mapsto f(x) \in f(K) = f(A)$ is an isomorphism.

Suppose that $f(A) \not\subseteq pA$ whenever $pA \neq A$. It suffices to show that $f(A)$ contains a non–zero direct summand of A for the following reason. If $0 \neq L \subseteq f(K)$ and $L \subseteq^{\oplus} A$, then $f^{-1}(L) \subseteq^{\oplus} K \subseteq^{\oplus} A$, so $f^{-1}(L) \subseteq^{\oplus} A$ and $f^{-1}(L) \xrightarrow{f} L$ is an isomorphism, i.e., f is pi.

The purification $f(A)^A_*$ is a direct summand of A (Proposition 1.27) and hence τ–homogeneous completely decomposable, as is $f(A) = f(K) \cong K$. The quotient $f(A)^A_*/f(A)$ is a torsion group which implies that $\operatorname{rk}(f(A)) = \operatorname{rk}(f(A)^A_*)$. This says that $f(A)^A_* \cong f(A)$. By Proposition 1.28 the quotient $f(A)^A_*/f(A)$ must actually be finite. Let $m := \exp(f(A)^A_*/f(A))$. We are looking for an element $x \in f(A)$ with the property that $\mathbb{Q}^A_x = \mathbb{Q}^{f(A)}_x$; for such an element $\mathbb{Q}^A_x x = \mathbb{Q}^{f(A)}_x x \subseteq f(A)$ and $\mathbb{Q}^A_x x$ is pure in A and therefore a direct summand. It is immediate that $\mathbb{Q}^{f(A)}_x \subseteq \mathbb{Q}^A_x$ for any $x \in f(A)$. For the converse containment it suffices to show that $1/p^t \in \mathbb{Q}^A_x$ implies that $1/p^t \in \mathbb{Q}^{f(A)}_x$ for any prime p. We first observe that for any $x \in f(A)$ and any p that is not a divisor of m, the implication $1/p^t \in \mathbb{Q}^A_x \Rightarrow 1/p^t \in \mathbb{Q}^{f(A)}_x$ is valid. It remains to consider the prime divisors $p_1, \ldots, p_k$ of m and to find x. For each p_i there is an element $x_i \in f(A)$ such that $x_i \notin p_i A$ or, equivalently, $1/p_i \notin \mathbb{Q}^A_{x_i}$. Let

$$\begin{aligned} x &= (p_2 \cdots p_k)x_1 + (p_1 p_3 \cdots p_k)x_2 + \cdots \\ &\quad + (p_1 p_2 \cdots p_{i-1} p_{i+1} \cdots p_k)x_i + \cdots + (p_1 \cdots p_{k-1})x_k. \end{aligned}$$

Then for each i, $1/p_i \notin \mathbb{Q}^A_x$ since $1/p_i \in \mathbb{Q}^A_{(p_1 p_2 \cdots p_{j-1} p_{j+1} \cdots p_k)x_j}$ for $j \neq i$ and $1/p_i \notin \mathbb{Q}^A_{(p_1 p_2 \cdots p_{i-1} p_{i+1} \cdots p_k)x_i}$. This completes the proof that $\mathbb{Q}^{f(A)}_x = \mathbb{Q}^A_x$ and establishes the claim as explained above.

The description of the total follows from the observation that an endomorphism f of A with $f(A) \subseteq pA$ can be factored as $f = p(p^{-1}f)$ with $p^{-1}f \in \operatorname{End}(A)$ while the converse is trivial. □

Corollary 2.2. *Let A be a completely decomposable homogeneous group of finite rank. Then* $\operatorname{Tot}(\operatorname{End}(A)) = 0$ *if and only if A is a divisible group.*

Proof. Let p be a prime. Multiplication by p is an endomorphism of A with image pA. By Theorem 2.1 we must have $A = pA$ or else p is a non–zero element of the total of $\operatorname{End}(A)$. Hence A is p–divisible for each p, so divisible. Conversely, a divisible torsion–free Abelian group of finite rank is a $\mathbb{Q}$–vector space of finite dimension. In this case the total of $\operatorname{End}(A)$ coincides with the radical of $\operatorname{End}(A)$ which is 0. □

For the case of general completely decomposable groups we need a number of lemmas.

Lemma 2.3. *Suppose that $B \subseteq^{\oplus} A$, and $f \in \operatorname{End}(A)$ is such that $f(B) \subseteq B$ and $f\restriction_B$ is partially invertible in* $\operatorname{End}(B)$. *Then f is partially invertible in* $\operatorname{End}(A)$. *Equivalently, if $f \in \operatorname{Tot}(\operatorname{End}(A))$, then $f\restriction_B \in \operatorname{Tot}(\operatorname{End}(B))$.*

Proof. By hypothesis there are decompositions $B = B_1 \oplus C_1 = B_2 \oplus C_2$ such that $B_1 \ni x \mapsto f(x) \in B_2$ is an isomorphism. Then the B_i are also summands of A and f is pi in A. □

Lemma 2.4. *Let M be an R–module and L a fully invariant submodule. Let $\nu : M \to M/L$ be the natural epimorphism. Then there are two ring homomorphisms $\rho : \operatorname{End}(M) \ni f \mapsto f\restriction_L \in \operatorname{End}(L)$ and $\sigma : \operatorname{End}(M) \to \operatorname{End}(M/L)$ given by $\nu f = \sigma(f)\nu$ where $f \in \operatorname{End}(M)$. If f is partially invertible in $\operatorname{End}(M)$, then $\rho(f)$ is partially invertible in $\operatorname{End}(L)$ or $\sigma(f)$ is partially invertible in $\operatorname{End}(M/L)$. Equivalently, if $\rho(f) \in \operatorname{Tot}(\operatorname{End}(L))$ and $\sigma(f) \in \operatorname{Tot}(\operatorname{End}(M/L))$, then $f \in \operatorname{Tot}(\operatorname{End}(M))$.*

Proof. Let f be pi in $\operatorname{End}(M)$. Then $fg = e = e^2 \neq 0$ in $\operatorname{End}(M)$, and hence $\rho(f)\rho(g) = \rho(e) = \rho(e)^2$ in $\operatorname{End}(L)$. So $\rho(f)$ is pi in $\operatorname{End}(L)$ provided that $\rho(e) \neq 0$. Suppose that $\rho(e) = 0$. Then $e = d\nu$ for some $d \in \operatorname{Hom}(M/L, M)$. If $\sigma(e) = 0$, then $e(M) = d(M) \subseteq L$ and we obtain the contradiction $e = e^2 = (d\nu)(d\nu) = d(\nu d)\nu = d0\nu = 0$. So $\sigma(f)\sigma(g) = \sigma(e) = \sigma(e)^2 \neq 0$ showing that $\sigma(f)$ is pi in $\operatorname{End}(M/L)$. □

The third lemma essentially combines the previous two.

Lemma 2.5. *Let M be an R–module and suppose that L is a fully invariant direct summand of M. Let $\rho : \operatorname{End}(M) \to \operatorname{End}(L)$ be the restriction map. Then*

$$\rho(\operatorname{Tot}(\operatorname{End}(M))) = \operatorname{Tot}(\operatorname{End}(L)).$$

Proof. By Lemma 2.3 we have $\rho(\operatorname{Tot}(\operatorname{End}(M))) \subseteq \operatorname{Tot}(\operatorname{End}(L))$.

Let $\sigma : \operatorname{End}(M) \to \operatorname{End}(M/L)$ be as in Lemma 2.4. Since L is a direct summand of M there is a mapping $\lambda : \operatorname{End}(L) \to \operatorname{End}(M)$ such that $\rho\lambda = 1_{\operatorname{End}(L)}$ and $\sigma\lambda = 0_{\operatorname{End}(M/L)}$. So given $f \in \operatorname{End}(L)$ we obtain $\lambda(f) \in \operatorname{End}(M)$ with $\rho(\lambda(f)) = f$ and $\sigma(\lambda(f)) = 0$. It suffices to show that $f \in \operatorname{Tot}(\operatorname{End}(L))$ implies that $\lambda(f) \in \mathrm{Tot}(\mathrm{End}(M))$, or equivalently, if $\lambda(f)$ is pi in $\mathrm{End}(M)$, then f is pi in $\mathrm{End}(L)$. Suppose that indeed $\lambda(f)$ is pi in $\mathrm{End}(M)$. Then, by Lemma 2.4 either $f = \rho\lambda(f)$ is pi in $\mathrm{End}(L)$ or $\sigma\lambda(f)$ is pi in $\mathrm{End}(M/L)$. But $\sigma\lambda(f) = 0$ surely is not pi, and we conclude that f must be pi as desired. □

Since the type subgroups $G(\tau)$ and $G^\sharp(\tau)$ are fully invariant subgroups for any torsion–free group G, every endomorphism f of G induces an endomorphism f° of the quotient

$$G^\circ(\tau) := G(\tau)/G^\sharp(\tau).$$

We begin with the special case that the completely decomposable group A contains exactly one minimal critical type τ so that $A = A(\tau)$.

Proposition 2.6. *Let A be a completely decomposable group. Suppose that $A = A(\tau) = A_\tau \oplus A^\sharp(\tau)$ and $f \in \mathrm{End}(A(\tau))$. Then f is partially invertible if and only if $f \restriction_{A^\sharp(\tau)}$ is partially invertible or the induced map $f^\circ \in \mathrm{End}\big(A(\tau)/A^\sharp(\tau)\big)$ is partially invertible.*

Proof. Suppose that f is pi in $\mathrm{End}(A)$. Then there exist decompositions $A = B \oplus C = D \oplus E$ such that $B \neq 0$ and $B \ni b \mapsto f(b) \in D$ is an isomorphism. We also have decompositions $A^\sharp(\tau) = B^\sharp(\tau) \oplus C^\sharp(\tau) = D^\sharp(\tau) \oplus E^\sharp(\tau)$ and decompositions $B = B_\tau \oplus B^\sharp(\tau)$, $C = C_\tau \oplus C^\sharp(\tau)$, $D = D_\tau \oplus D^\sharp(\tau)$, $E = E_\tau \oplus E^\sharp(\tau)$. We obtain the induced map $B^\sharp(\tau) \ni b \mapsto f(b) \in D^\sharp(\tau)$ which is surely injective. The identity

$$D_\tau \oplus D^\sharp(\tau) = D = f(B) = f(B_\tau) \oplus f(B^\sharp(\tau))$$

with $f(B_\tau) \cong B_\tau$ and $f(B^\sharp(\tau)) \subseteq D^\sharp(\tau)$ implies that $f(B^\sharp(\tau)) = D^\sharp(\tau)$. Hence $B^\sharp(\tau) \ni b \to f(b) \in D^\sharp(\tau)$ is an isomorphism and $f \restriction_{A^\sharp(\tau)}$ is pi provided that $B^\sharp(\tau) \neq 0$.

Assume now that $B^\sharp(\tau) = 0$. Then $D^\sharp(\tau) = 0$ as well, and we have that $A^\sharp(\tau) = C^\sharp(\tau) = E^\sharp(\tau)$ and $A = B \oplus C_\tau \oplus A^\sharp(\tau) = D \oplus E_\tau \oplus A^\sharp(\tau)$. We obtain decompositions

$$\frac{A(\tau)}{A^\sharp(\tau)} = \frac{B \oplus A^\sharp(\tau)}{A^\sharp(\tau)} \oplus \frac{C_\tau \oplus A^\sharp(\tau)}{A^\sharp(\tau)} = \frac{D \oplus A^\sharp(\tau)}{A^\sharp(\tau)} \oplus \frac{E_\tau \oplus A^\sharp(\tau)}{A^\sharp(\tau)}$$

and isomorphisms

$$\begin{array}{ccccc} \frac{B\oplus A^\sharp(\tau)}{A^\sharp(\tau)} & \ni b + A^\sharp(\tau) & \overset{f^\circ}{\mapsto} & f(b) + A^\sharp(\tau) \in & \frac{D\oplus A^\sharp(\tau)}{A^\sharp(\tau)} \\ \downarrow & & & & \downarrow \\ B & \ni b & \mapsto & f(b) \in & C \end{array}.$$

This shows that f° is pi.

Conversely, if $f \restriction_{A^\sharp(\tau)}$ is pi, then f is pi by Lemma 2.3. This leaves the case that $f^\circ \in \mathrm{End}\big(A(\tau)/A^\sharp(\tau)\big)$ is pi. In this case there exist decompositions

$$\frac{A(\tau)}{A^\sharp(\tau)} = \frac{B}{A^\sharp(\tau)} \oplus \frac{C}{A^\sharp(\tau)} = \frac{D}{A^\sharp(\tau)} \oplus \frac{E}{A^\sharp(\tau)}$$

such that

$$\frac{B}{A^\sharp(\tau)} \ni b + A^\sharp(\tau) \mapsto f(b) + A^\sharp(\tau) \in \frac{D}{A^\sharp(\tau)} \tag{2}$$

is an isomorphism. We also have decompositions $B = B_\tau \oplus A^\sharp(\tau)$ and $D = D_\tau \oplus A^\sharp(\tau)$. It is easy to see that $f(B) \subseteq D$. Let $y \in D_\tau$. Then there is $x \in B_\tau$ such that $f(x) + A^\sharp(\tau) = y + A^\sharp(\tau)$, hence $f(x) = y + z$ for some $z \in A^\sharp(\tau)$. Given $y \in D_\tau$, the element z is uniquely determined, hence we have a well–defined map

$$f : D_\tau \to A^\sharp(\tau) : f(y) = z \quad \text{where} \quad f(x) = y + z, x \in B_\tau.$$

It is straightforward to check that f is a homomorphism. Hence $D = (f+1)D_\tau \oplus A^\sharp(\tau)$ and we have the isomorphism

$$B_\tau \ni x \mapsto f(x) \in (f+1)D_\tau$$

showing that f is pi in $\operatorname{End}(A)$. □

Remark. By inspecting the proof of Proposition 2.6 one realizes that it was not used that A was a completely decomposable group but what was needed were the decompositions $B = B_\tau \oplus B^\sharp(\tau)$, $C = C_\tau \oplus C^\sharp(\tau)$, $D = D_\tau \oplus D^\sharp(\tau)$, $E = E_\tau \oplus E^\sharp(\tau)$ for the direct summands B, C, D, E of A. Such decompositions are known as Butler–decompositions and they do exist for the so–called Butler groups, i.e., the pure subgroups of completely decomposable groups. Thus Proposition 2.6 is true for a Butler group A.

A completely decomposable group of finite rank whose total is trivial must be very special. The following theorem generalizes Corollary 2.2.

Theorem 2.7. *Let A be a completely decomposable group of finite rank. Then* $\operatorname{Tot}(\operatorname{End}(A)) = 0$ *if and only if A is a divisible group.*

For the proof we need a lemma.

Lemma 2.8. *Let A be any torsion–free Abelian group. Then for any $n \in \mathbb{N}$ there is a unique largest n–divisible subgroup D_n of A. Furthermore, D_n is pure in A.*

Proof. Let $\mathcal{U}$ be the family of all n–divisible subgroups of A. Then $D_n := \sum \mathcal{U}$ is also an n–divisible subgroup of A, and hence the unique largest such group.

Suppose that $a \in A$ and $ma \in D_n$ for some non–zero integer m. Factor m such that $m = m_1 m_2$, $\gcd(m_1, n) = 1$ and every prime divisor of m_2 is a prime divisor of n. Then $m_1 a = \frac{1}{m_2}(ma) \in D_n$. So without loss of generality $m = m_1$. For any positive integer k, there exist integers u, v such that $1 = um + vn^k$. It follows that $\frac{1}{n^k}a = u\frac{1}{n^k}(ma) + va \in A$. Hence $\{sn^{-k}a \mid s \in \mathbb{Z}, k \in \mathbb{N}_0\}$ is an n–divisible subgroup of A and hence contained in D_n. In particular, $a \in D_n$, showing that D_n is pure in A. □

Proof of Theorem 2.7. Let n be a positive integer. Because $\mathrm{Tot}(\mathrm{End}(A)) = 0$ by assumption, multiplication by n in A is pi, and therefore there exists a non–zero direct summand B of A such that $nB = B$. Hence A contains n–divisible subgroups for any n. Consequently, there is a descending chain of non–zero subgroups

$$D_{2!} \supseteq D_{3!} \supseteq \cdots \supseteq D_{n!} \supseteq D_{(n+1)!} \cdots$$

where $D_{n!}$ denotes the largest $n!$–divisible subgroup of A (Lemma 2.8). Among the groups in the chain there must be one of least rank, say $D_{N!}$. The fact that all the groups of the chain are pure now says that $D_{N!} = D_{(N+1)!} = \cdots$ and this means that $D_{N!}$ is divisible. We have found that A possesses direct summands isomorphic with $\mathbb{Q}$. We will show that every rank–one summand of A is isomorphic with $\mathbb{Q}$ and then an application of Corollary 2.2 finishes the proof. Suppose to the contrary that A has summands not isomorphic with $\mathbb{Q}$. Then there is a decomposition $A = Bu \oplus \mathbb{Q}v \oplus A'$ where B is a rational group not equal to $\mathbb{Q}$. Let f be the endomorphism of A given by $f(u) = v$ and $f\restriction_{\mathbb{Q}v\oplus A'} = 0$. Then f must be pi as $\mathrm{Tot}(\mathrm{End}(A)) = 0$, hence there exist non–zero direct summands C_1 and C_2 of A such that $C_1 \ni c \mapsto f(c) \in C_2$ is an isomorphism. Now $C_2 \subseteq \mathrm{Im}(f) \subseteq \mathbb{Q}v$ and C_2 is pure, so $C_2 = \mathbb{Q}v \cong \mathbb{Q}$. Also $C_2 \cong C_1 \cong (C_1 \oplus \mathrm{Ker}(f))/\mathrm{Ker}(f)) \subseteq (Bu \oplus \mathrm{Ker}(f))/\mathrm{Ker}(f) \cong B$ showing that C_2 cannot be divisible. This is the desired contradiction. □

We now generalize the special case Proposition 2.6 to arbitrary completely decomposable groups. To do so, we need to generalize the set–up. The proof will then be essentially as in the special case. In Proposition 2.6 it was assumed that $\mathrm{T_{cr}}(A)$ contains a single minimal element. In general, there will be several minimal elements.

Definition 2.9. Let $A = \bigoplus_{\rho\in\mathrm{T_{cr}}(A)} A_\rho$ be the homogeneous decomposition of the completely decomposable group A. Set

$$\mathrm{T_{mcr}}(A) := \{\rho \in \mathrm{T_{cr}} \mid \rho \text{ is minimal in } \mathrm{T_{cr}}(A)\}$$

and let

$$A_{min} := \bigoplus_{\rho\in\mathrm{T_{mcr}}(A)} A_\rho, \quad A^\sharp := \bigoplus_{\rho\notin\mathrm{T_{mcr}}(A)} A_\rho.$$

The essential relationships of $A(\tau)$ and $A^\sharp(\tau)$ used in the proof of Proposition 2.6 are still true for A and $A^\sharp$ as we will show next. However, it is essential for the validity of Lemma 2.10 that A is completely decomposable so that Proposition 2.6 is not simply a special case of Theorem 2.11.

Lemma 2.10. *Let $A = \bigoplus_{\rho\in\mathrm{T_{cr}}(A)} A_\rho$ be the homogeneous decomposition of the completely decomposable group A. Then the following statements hold.*

1) $A = A_{min} \oplus A^\sharp$.

2) *$A^\sharp$ is fully invariant in A, in fact, $A^\sharp = \sum_{\rho\in\mathrm{T_{cr}}(A)} A^\sharp(\rho)$.*

3) *If* $A = B \oplus C$, *then* $B^\sharp = B \cap A^\sharp$, $C^\sharp = C \cap A^\sharp$, *and* $A^\sharp = B^\sharp \oplus C^\sharp$.

4) *An isomorphism* $f : B \to D$ *between completely decomposable groups* B, D *induces by restriction the isomorphism* $B^\sharp \ni b \mapsto f(b) \in C^\sharp$.

Proof. 1) The decomposition $A = A_{min} \oplus A^\sharp$ is obtained by collecting homogeneous components according to the definition.

2) Both $A^\sharp$ and $\sum_{\rho \in \mathrm{T_{cr}}(A)} A^\sharp(\rho)$ are sums of the same homogeneous components A_ρ. The sum $\sum_{\rho \in \mathrm{T_{cr}}(A)} A^\sharp(\rho)$ is fully invariant as a sum of fully invariant subgroups.

3) By 2) $A^\sharp$ is fully invariant in A and therefore

$$A^\sharp = A^\sharp \cap A = A^\sharp \cap B \oplus A^\sharp \cap C. \tag{3}$$

By Theorem 1.26 the summands B, C are again completely decomposable. We have $B = \bigoplus_{\rho \in \mathrm{T_{cr}}(B)} B_\rho$ and $C = \bigoplus_{\rho \in \mathrm{T_{cr}}(C)} C_\rho$. Thus

$$A = \left(\bigoplus\nolimits_{\rho \in \mathrm{T_{cr}}(B)} B_\rho\right) \oplus \left(\bigoplus\nolimits_{\rho \in \mathrm{T_{cr}}(C)} C_\rho\right),$$

and it follows that

$$\begin{aligned} A^\sharp &= \left(\bigoplus\nolimits_{\rho \notin \mathrm{T_{mcr}}(B)} B_\rho\right) \oplus \left(\bigoplus\nolimits_{\rho \notin \mathrm{T_{mcr}}(C)} C_\rho\right) \\ &= B^\sharp \oplus C^\sharp. \end{aligned} \tag{4}$$

Evidently $B^\sharp \subseteq B \cap A^\sharp$ and $C^\sharp \subseteq C \cap A^\sharp$. The decompositions (3) and (4) together imply that $B^\sharp = B \cap A^\sharp$ and $C^\sharp = C \cap A^\sharp$.

4) We note that $\mathrm{T_{cr}}(B) = \mathrm{T_{cr}}(C)$ and so also $\mathrm{T_{mcr}}(B) = \mathrm{T_{mcr}}(C)$. By 1) it follows that $f(B^\sharp) \subseteq C^\sharp$ and $f^{-1}(C^\sharp) \subseteq B^\sharp$. We conclude that $f(B^\sharp) = C^\sharp$ and the restriction of f is automatically injective. □

Theorem 2.11. *Let* A *be a completely decomposable group. Then* $f \in \mathrm{End}(A)$ *is partially invertible if and only if* $f \restriction_{A^\sharp}$ *is partially invertible or the induced map* $f^\circ \in \mathrm{End}\left(A/A^\sharp\right)$ *is partially invertible.*

Proof. Suppose that f is pi in $\mathrm{End}(A)$. Then there exist decompositions $A = B \oplus C = D \oplus E$ such that $B \neq 0$ and $B \ni b \mapsto f(b) \in D$ is an isomorphism. As direct summands of a completely decomposable group, the groups B, C, D, E are again completely decomposable. We have decompositions $A^\sharp = B^\sharp \oplus C^\sharp = D^\sharp \oplus E^\sharp$ and decompositions $B = B_{min} \oplus B^\sharp$, $C = C_{min} \oplus C^\sharp$. The induced map $B^\sharp \ni b \mapsto f(b) \in D^\sharp$ is an isomorphism (Lemma 2.10.4) and $f \restriction_{A^\sharp}$ is pi provided that $B^\sharp \neq 0$.

Assume now that $B^\sharp = 0$. Then $D^\sharp = 0$ as well, and we have that $A^\sharp = C^\sharp = E^\sharp$ and $A = B \oplus C_{min} \oplus A^\sharp = D \oplus E_{min} \oplus A^\sharp$. We obtain decompositions

$$\frac{A}{A^\sharp} = \frac{B \oplus A^\sharp}{A^\sharp} \oplus \frac{C_{min} \oplus A^\sharp}{A^\sharp} = \frac{D \oplus A^\sharp}{A^\sharp} \oplus \frac{E_{min} \oplus A^\sharp}{A^\sharp}$$

and isomorphisms

$$\begin{array}{ccccc} \frac{B\oplus A^\sharp}{A^\sharp} & \ni b + A^\sharp & \overset{f^\circ}{\mapsto} & f(b) + A^\sharp \in & \frac{D\oplus A^\sharp}{A^\sharp} \\ \downarrow & & & & \downarrow \\ B & \ni b & \mapsto & f(b) \in & D \end{array}.$$

This shows that f° is pi.

Conversely, if $f\restriction_{A^\sharp(\tau)}$ is pi, then f is pi by Lemma 2.3. This leaves the case that $f^\circ \in \mathrm{End}\left(A/A^\sharp\right)$ is pi. In this case there exist decompositions

$$\frac{A}{A^\sharp} = \frac{B}{A^\sharp} \oplus \frac{C}{A^\sharp} = \frac{D}{A^\sharp} \oplus \frac{E}{A^\sharp}$$

such that

$$\frac{B}{A^\sharp} \ni b + A^\sharp \mapsto f(b) + A^\sharp \in \frac{D}{A^\sharp} \tag{5}$$

is an isomorphism. We also have decompositions $B = B_{min} \oplus A^\sharp$ and $D = D_{min} \oplus A^\sharp$. It is evident that $f(B) \subseteq D$. Let $y \in D_{min}$. Then there is $x \in B_{min}$ such that $f(x) + A^\sharp = y + A^\sharp$. If also $x' \in B_{min}$ and $f(x') + A^\sharp = y + A^\sharp$, then, (5) being injective, $x' + A^\sharp = x + A^\sharp$ and further $x' - x \in B_{\min} \cap A^\sharp = B_{min} \cap B^\sharp = 0$. So x is unique. Hence, given $y \in D_{min}$, the element $z = f(x) - y \in A^\sharp$ is uniquely determined and we have a well–defined map

$$f : D_{min} \to A^\sharp : f(y) = z \quad \text{where} \quad f(x) = y + z,\ x \in B_{min}.$$

It is straightforward to check that f is a homomorphism. Hence $D = (f+1)D_{min} \oplus A^\sharp$ and we have the isomorphism

$$B_{min} \ni x \mapsto f(x) \in (f+1)D_{min}$$

showing that f is pi in $\mathrm{End}(A)$. □

By negation we have the following corollary.

Corollary 2.12. *Let A be a completely decomposable group. Then $f \in \mathrm{Tot}(\mathrm{End}(A))$ if and only if*

$$f\restriction_A^\sharp \in \mathrm{Tot}(\mathrm{End}(A^\sharp)) \text{ and } f^\circ \in \mathrm{Tot}(\mathrm{End}(A/A^\sharp)).$$

The preceding Corollary 2.12 is looked at as a reduction theorem. The pi elements of the endomorphism ring of the completely decomposable group $A^\sharp$ may be considered known (by an induction on the depth of the critical typeset, see below) and we will consider next the pi elements of the endomorphism ring of A_{min}. Recall that $A/A^\sharp \cong A_{min} = \bigoplus_{\rho\in \mathrm{T}_{\mathrm{mcr}}(A)} A_\rho$. The minimal types in $\mathrm{T}_{\mathrm{cr}}(A)$ are pairwise incomparable and homomorphisms can never decrease types (Lemma 1.15.5)). For this reason we obtain a ring direct product

$$\mathrm{End}(A_{min}) = \prod\nolimits_{\rho\in \mathrm{T}_{\mathrm{mcr}}(A)} \mathrm{End}(A_\rho).$$

The total of the endomorphism ring of a completely decomposable group $A = \bigoplus_{\rho\in \mathrm{T}_{\mathrm{cr}}(A)} A_\rho$ is now determined in the following sense.

1) Write $A = A_{min} \oplus A^{\sharp}$.

2) Write $A_{min} = \bigoplus_{\rho \in \mathrm{T}_{\mathrm{mcr}}(A)} A_\rho \cong A/A^{\sharp}$.

3) Find the total of $\mathrm{End}(A_\mu)$ for $\mu \in \mathrm{T}_{\mathrm{mcr}}(A)$ using Theorem 2.1.

4) Find the total of $\mathrm{End}(A_{min})$ using Proposition 1.8.

5) If $\mathrm{T}_{\mathrm{cr}}(A) \setminus \{\mu \mid \mu \in \mathrm{T}_{\mathrm{mcr}}(A)\}$ is an anti-chain, i.e., any two of these types are incomparable, then Theorem 2.1 applies to $\mathrm{End}(A^{\sharp})$ and Corollary 2.12 can be applied to determine the functions of $\mathrm{Tot}(\mathrm{End}(A))$.

6) In any case, the critical type set

$$\mathrm{T}_{\mathrm{cr}}(A^{\sharp}) = \{\rho \in \mathrm{T}_{\mathrm{cr}}(A) \mid \rho \notin \mathrm{T}_{\mathrm{mcr}}(A)\}$$

has one layer of minimal types less than $\mathrm{T}_{\mathrm{cr}}(A)$ and by the procedure described here the total of $\mathrm{End}(A^{\sharp})$ can be obtained recursively.

3 A category of torsion–free Abelian groups with LE–decompositions

In group and module theory there is an interest into decompositions into direct sums of indecomposable subobjects. For convenience we call a direct decomposition an **indecomposable decomposition** if the direct summands are all indecomposable. In general, torsion–free Abelian groups are notorious for their essentially different indecomposable decompositions. The main point of this section is to demonstrate that the lack of uniqueness can be salvaged at the expense of introducing an equivalence of groups that is weaker than isomorphism. This leads to the so–called "quasi–isomorphism category". In this category the total of an endomorphism ring coincides with the radical of the endomorphism ring. Some relationships between the totals of the ordinary endomorphism ring of a group and its quasi–endomorphism ring are noted.

We begin with an example, that may be considered folklore, of a group X that has two essentially different indecomposable decompositions. The group X is a **finite essential extension** of a completely decomposable group C, meaning that C is large in X. This is equivalent to saying that X is also torsion–free. The group C has a homogeneous direct summand of rank 2 that can be decomposed in different ways and as a consequence the elements of X ("clamps") that tie the summands of C together are replaced by clamps that tie together different summands of C. The group may be depicted as follows. The homogeneous block of rank 2 pictured on the left is decomposed in two different ways which replaces a single 3–pronged clamp by two 2–pronged clamps.

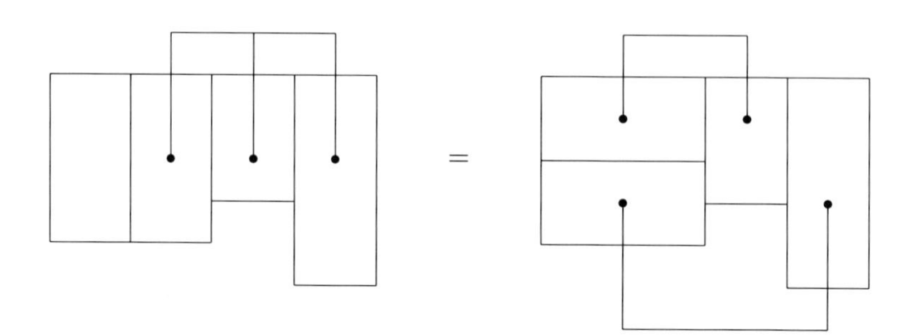

Example 3.1. Let V be a 4-dimensional $\mathbb{Q}$-vector space with basis $\{v_1, v_2, v_3, v_4\}$. Then

$$V = \mathbb{Q}v_1 \oplus \mathbb{Q}v_2 \oplus \mathbb{Q}v_3 \oplus \mathbb{Q}v_4.$$

The example will be obtained as an additive subgroup of V. Let $A := \mathbb{Z}[5^{-1}]$, $A_1 := \mathbb{Z}[7^{-1}]$, $A_2 := \mathbb{Z}[11^{-1}]$,

$$C = Av_1 \oplus Av_2 \oplus A_1v_3 \oplus A_2v_4, \text{ and } X = C + \mathbb{Z}\frac{1}{2 \cdot 3}(v_2 + 3v_3 + 2v_4).$$

Then

$$\begin{aligned} X &= Av_1 \oplus \left((Av_2 \oplus A_1v_3 \oplus A_2v_4) + \mathbb{Z}\frac{1}{2 \cdot 3}(v_2 + 3v_3 + 2v_4) \right) \\ &= \left((A(2v_1 + v_2) \oplus A_1v_3) + \mathbb{Z}\frac{1}{2}(2v_1 + v_2 + 3v_3) \right) \\ &\quad \oplus \left((A(3v_1 + v_2) \oplus A_2v_4) + \mathbb{Z}\frac{1}{3}(3v_1 + v_2 + 2v_4) \right) \end{aligned}$$

are indecomposable decompositions with summands of rank 1 and 3 in the first decomposition and two summands of rank 2 in the second decomposition.

Proof. We first observe that the two decompositions

$$Av_1 \oplus Av_2 = A(2v_1 + v_2) \oplus A(3v_1 + v_2)$$

are correct. This can be checked directly but it is perhaps more conceptual to observe that the decomposition on the right-hand side arises from the decomposition on the left by applying the automorphism given by left multiplication by an integral matrix of determinant -1:

$$\begin{bmatrix} 2 & 1 \\ 3 & 1 \end{bmatrix} \begin{bmatrix} v_1 \\ v_2 \end{bmatrix} = \begin{bmatrix} 2v_1 + v_2 \\ 3v_1 + v_2 \end{bmatrix}.$$

The correctness of the decompositions of X follows from the identities

$$\begin{aligned}\frac{1}{2}(2v_1+v_2+3v_3) &= 3\frac{1}{6}(v_2+3v_3+2v_4)+v_1-v_4 \in X,\\ \frac{1}{3}(3v_1+v_2+2v_4) &= 2\frac{1}{6}(v_2+3v_3+2v_4)+v_1-v_3 \in X,\\ \frac{1}{6}(v_2+3v_3+2v_4) &= \frac{1}{2}(2v_1+v_2+3v_3)-\frac{1}{3}(3v_1+v_2+2v_4)-v_3+v_4.\end{aligned}$$

To show that the summands are indecomposable one can check that their endomorphism rings contain no non–trivial idempotents. As an example consider

$$Y := (Av_2 \oplus A_1v_3 \oplus A_2v_4)+\mathbb{Z}\frac{1}{2\cdot 3}(v_2+3v_3+2v_4).$$

One must first recognize that the subgroup $B := Av_2 \oplus A_1v_3 \oplus A_2v_4$ is fully invariant in Y and that the restriction of an endomorphism φ of Y is completely determined by its action on B. Now A is rigid, meaning that there are no non–zero homomorphisms between any two of the summands Av_2, A_1v_3 and A_2v_4. This means that the endomorphism ring of B can be identified with $A \times A_1 \times A_2$ where the action of a triple (r, r_1, r_2), $r \in A$, $r_1 \in A_1$, $r_2 \in A_2$ is given by

$$(r, r_1, r_2)(sv_2+s_1v_3+s_2v_4) = rsv_2+r_1s_1v_3+r_2s_2v_4.$$

The endomorphism ring of Y consists of all triples (r, r_1, r_2) with the property that $\frac{1}{2\cdot 3}(rv_2+3r_1v_3+2r_2v_4) \in Y$. The idempotents of $\mathrm{End}(B)$ obviously are the eight maps $(0,0,0)$, $(1,0,0)$, $(0,1,0)$, $(0,0,1)$, $(1,1,0)$, $(1,0,1)$, $(0,1,1)$, $(1,1,1)$, and an idempotent in $\mathrm{End}(Y)$ restricts to an idempotent of $\mathrm{End}(A)$ and therefore must be one of the eight idempotents listed. Suppose, e.g., that $(1,0,1) \in \mathrm{End}(Y)$. Then

$$\frac{1}{6}(v_2+2v_4) = m\frac{1}{6}(v_2+3v_3+2v_4)+sv_2+s_1v_3+s_2v_4,$$

for elements $m \in \mathbb{Z}$, $s \in A$, $s_1 \in A_1$, $s_2 \in A_2$. Hence $1 = m+6s$, $0 = 3m+6s_1$, $2 = 2m+6s_2$, from which a contradiction is easily derived. All but the trivial idempotents $(0,0,0)$ and $(1,1,1)$ can be eliminated in this fashion. □

A very striking and easily stated result on "pathological decomposition" is as follows.

Theorem 3.2. (Corner [10], or [25, p.281]) *Given integers $n \geq k \geq 1$, there exists a torsion–free group X of rank n such that for any partition $n = r_1+\cdots+r_k$, there is a decomposition of X into a direct sum of k indecomposable subgroups of ranks $r_1, \ldots, r_k$ respectively.*

Let $\mathcal{A}$ denote the category of torsion–free Abelian groups of finite rank with the usual homomorphisms $\mathrm{Hom}(G, H)$ as morphisms. The problem with "pathological decompositions" can be remedied by changing the category as follows. This idea goes back to Bjarni Jonsson ([13], [14]).

Let G and H be groups in $\mathcal{A}$. Recall Lemma 1.3 which can be interpreted to mean that

$$\operatorname{Hom}(G,H) \subseteq \operatorname{Hom}_{\mathbb{Q}}(\mathbb{Q}G,\mathbb{Q}H) = \operatorname{Hom}(\mathbb{Q}G,\mathbb{Q}H).$$

Let $\mathbb{Q}\operatorname{Hom}(G,H) = \{rf \mid r \in \mathbb{Q}, f \in \operatorname{Hom}(G,H)\}$ be the subspace of $\operatorname{Hom}(\mathbb{Q}G,\mathbb{Q}H)$ generated by $\operatorname{Hom}(G,H)$. In particular, $\mathbb{Q}\operatorname{Hom}(G,H)$ is an additive Abelian group with the addition inherited from the group of linear mappings $\operatorname{Hom}(\mathbb{Q}G,\mathbb{Q}H)$. If G, H, K are three groups in $\mathcal{A}$, $f \in \mathbb{Q}\operatorname{Hom}(G,H) \subseteq \operatorname{Hom}(\mathbb{Q}G,\mathbb{Q}H)$, and $g \in \mathbb{Q}\operatorname{Hom}(H,K) \subseteq \operatorname{Hom}(\mathbb{Q}H,\mathbb{Q}K)$, then the composite linear mapping gf is not only in $\operatorname{Hom}(\mathbb{Q}G,\mathbb{Q}H)$ but in $\mathbb{Q}\operatorname{Hom}(G,H)$. This says that the "category" defined next is indeed a category.

Definition 3.3. Let $\mathbb{Q}\mathcal{A}$ denote the category whose objects are those of $\mathcal{A}$, and whose morphism sets are the Abelian groups $\mathbb{Q}\operatorname{Hom}(G,H)$ with addition and composition inherited from $\operatorname{Hom}(\mathbb{Q}G,\mathbb{Q}H)$. This is the **quasi–isomorphism category** of torsion–free Abelian groups of finite rank.

Recall first that a **preadditive category** is a category $\mathcal{C}$ in which the morphism sets are Abelian groups with the property that composition of morphisms distributes over addition. It is also postulated that the category contain a null object whose endomorphism group is the zero group. An object A in a preadditive category $\mathcal{C}$ is a **biproduct** of the objects $A_1, \dots, A_n \in \mathcal{C}$ if there exist morphisms, called **structural maps**,

$$\iota_i : A_i \to A, \quad \text{and} \quad \pi_i : A \to A_i$$

such that

$$\pi_i\iota_i = 1_{A_i}, \quad \pi_j\iota_i = 0 \text{ for } i \neq j, \quad \text{and} \quad \iota_1\pi_1 + \cdots + \iota_n\pi_n = 1_A.$$

If this is the case we write $A \in A_1 \oplus \cdots \oplus A_n$, and think of $A_1 \oplus \cdots \oplus A_n$ as the collection of all data sets (A, ι_i, π_i). A biproduct has the universal properties of both products and co–products and is unique up to isomorphism in the category. If $A \in A_1 \oplus \cdots \oplus A_n$ and B is isomorphic with A, then $B \in A_1 \oplus \cdots \oplus A_n$ also. Further note that the apparent ordering of the summands is introduced by the choice of labels and is not intrinsic to the definition. An **additive category** is a preadditive category that has biproducts for any finite set of objects.

Proposition 3.4. *The quasi–isomorphism category is an additive category.*

Proof. The algebraic laws of the morphisms are automatic because the operations are inherited from vector space laws, each morphism set $\mathbb{Q}\operatorname{Hom}(G,H)$ is an Abelian group, and the group $\{0\}$ is the zero object of the category. Let $A_1, \dots A_n$ be in $\mathbb{Q}\mathcal{A}$. Then the ordinary biproduct $A_1 \oplus \cdots \oplus A_n$ in $\mathcal{A}$ is a biproduct in $\mathbb{Q}\mathcal{A}$. □

The following lemma contains the basic properties of the quasi–isomorphism category that are essential for working purposes. The routine verifications are left to the reader.

Lemma 3.5.

1) $\mathbb{Q}\operatorname{Hom}(G,H) = \{f \in \operatorname{Hom}(\mathbb{Q}G, \mathbb{Q}H) \mid \exists n \in \mathbb{N}, nf(G) \subseteq H\} = \{\frac{1}{n}f \mid n \in \mathbb{N}, f \in \operatorname{Hom}(G,H)\}$.

2) *Let* $f,g \in \mathbb{Q}\operatorname{Hom}(G,H)$. *Choose* $n \in \mathbb{N}$ *such that* $nf \in \operatorname{Hom}(G,H)$ *and* $ng \in \operatorname{Hom}(G,H)$. *Then* $f = g$ *in* $\mathbb{Q}\mathcal{A}$ *if and only if* $nf = ng$ *in* $\operatorname{Hom}(G,H)$.

3) *Let* $f,g \in \mathbb{Q}\operatorname{Hom}(G,H)$. *Choose* $n \in \mathbb{N}$ *such that* $nf \in \operatorname{Hom}(G,H)$ *and* $m \in \mathbb{N}$ *such that* $mg \in \operatorname{Hom}(G,H)$. *Then* $f + g = \frac{1}{nm}(nf + mg)$.

4) *Let* $f \in \mathbb{Q}\operatorname{Hom}(G,H)$ *and* $g \in \mathbb{Q}\operatorname{Hom}(H,K)$. *Choose* $n \in \mathbb{N}$ *such that* $nf \in \operatorname{Hom}(G,H)$ *and* $m \in \mathbb{N}$ *such that* $mg \in \operatorname{Hom}(H,K)$. *Then* $gf = \frac{1}{nm}((mg)(nf))$.

Two groups G,H that are isomorphic in the category $\mathbb{Q}\mathcal{A}$ are called **quasi–isomorphic** and we write $G \cong_{\text{qu}} H$. All $\mathbb{Q}\mathcal{A}$ notions have equivalent notions in $\mathcal{A}$. The interpretation of isomorphism in $\mathbb{Q}\mathcal{A}$ is as follows.

Proposition 3.6. *Let* $G,H \in \mathbb{Q}\mathcal{A}$. *Then the following statements are equivalent.*

1) $G \cong_{\text{qu}} H$,

2) *there exist* $0 \neq n \in \mathbb{N}$, $f \in \operatorname{Hom}(G,H)$, $g \in \operatorname{Hom}(H,G)$ *such that* $fg = n1_H$ *and* $gf = n1_G$,

3) *there exists a monomorphism* $f : G \to H$ *and* $0 \neq n \in \mathbb{N}$ *such that* $nH \subseteq f(G) \subseteq H$.

Proof. 1) $\Rightarrow$ 2) Suppose $G \cong_{\text{qu}} H$. Then there are $\frac{1}{k}f \in \mathbb{Q}\operatorname{Hom}(G,H)$, $\frac{1}{m}g \in \mathbb{Q}\operatorname{Hom}(H,G)$ such that $\frac{1}{k}f \cdot \frac{1}{m}g = 1_H$ and $\frac{1}{m}g \cdot \frac{1}{k}f = 1_G$. Hence $n = km$ will do.

2) $\Rightarrow$ 3) Suppose that $fg = n1_H$ and $gf = n1_G$, then f is injective and $nH = n1_H H = fg(H) \subseteq f(G) \subseteq H$.

3) $\Rightarrow$ 1) Suppose that $f : G \to H$ is a monomorphism and $0 \neq n \in \mathbb{N}$ such that $nH \subseteq f(G) \subseteq H$. Define $g : H \to G : g := f^{-1}n$. Then $\left(\frac{1}{n}f\right)g = 1_H$ and $g\left(\frac{1}{n}f\right) = 1_G$ hence $G \cong_{\text{qu}} H$. □

Example 3.7. The group X in Example 3.1 is quasi–isomorphic to its completely decomposable subgroup A: $X \cong_{\text{qu}} A$.

Proof. The inclusion of A in X is a monomorphism and the index $[X : A]$ is finite, in fact, a short computation shows that $[X : A] = 6$. □

We now interpret the meaning of biproduct in $\mathbb{Q}\mathcal{A}$ in the category $\mathcal{A}$. Since $H \oplus K$ has a meaning in $\mathcal{A}$, we will write $G \in H \oplus_{qu} K$ for the biproduct in $\mathbb{Q}\mathcal{A}$. The following example may serve as a warning.

Example 3.8. Let $G = \mathbb{Z}v_1 \oplus \mathbb{Z}v_2$ and $H = K = \mathbb{Z}(v_1 + v_2)$. Then $G \in H \oplus_{qu} K$ in $\mathbb{Q}\mathcal{A}$.

Proof. Let ι_H, π_H, ι_K, π_K be the following ordinary homomorphisms and so also quasi–homomorphisms.

- $\iota_H : H \ni \alpha(v_1 + v_2) \mapsto \alpha v_1 \in G$,
- $\pi_H : G \ni \alpha v_1 + \beta v_2 \mapsto \alpha(v_1 + v_2) \in H$,
- $\iota_K : K \ni \beta(v_1 + v_2) \mapsto \beta v_2 \in G$,
- $\pi_K : G \ni \alpha v_1 + \beta v_2 \mapsto \beta(v_1 + v_2) \in K$.

It is easily checked that these are well–defined mappings that satisfy the conditions required of structural maps. □

Proposition 3.9. *Let G, H, and K be torsion–free Abelian groups.*

1) *Let $nG \subseteq H \oplus K \subseteq G$ in $\mathcal{A}$ for some positive integer n. Then $G \in H \oplus_{qu} K$ in $\mathbb{Q}\mathcal{A}$. In particular, if $G = H \oplus K$ in $\mathcal{A}$, then $G \in H \oplus_{qu} K$.*

2) *Suppose that $G \in H \oplus_{qu} K$ in $\mathbb{Q}\mathcal{A}$. Then there are a positive integer n and subgroups H', K' of G such that $H' \cong H$, $K' \cong K$ and $nG \subseteq H' \oplus K' \subseteq G$.*

Proof. 1) By hypothesis and Proposition 3.6, $G \cong_{\mathrm{qu}} H \oplus K$, and hence $G \in H \oplus_{qu} K$.

2) Let $\iota_H \in \mathbb{Q}\mathrm{Hom}(H, G)$, $\pi_H \in \mathbb{Q}\mathrm{Hom}(G, H)$, $\iota_H \in \mathbb{Q}\mathrm{Hom}(H, G)$, and $\pi_H \in \mathbb{Q}\mathrm{Hom}(G, H)$ be the set of structural maps belonging to the quasi–decomposition $G \in H \oplus_{qu} K$. Then there is a positive integer m such that $m\iota_H \in \mathrm{Hom}(H, G)$, $m\pi_H \in \mathrm{Hom}(G, H)$, $m\iota_H \in \mathrm{Hom}(H, G)$, and $m\pi_H \in \mathrm{Hom}(G, H)$. Let $n = m^2$, $H' = (m\iota_H)(H)$ and $K' = (m\iota_K)(K)$. Then it follows from $\iota_H\pi_H + \iota_K\pi_K = 1_G$ that $(m\iota_H)(m\pi_H) + (m\iota_K)(m\pi_K) = m^2 1_G$, and therefore for every $x \in G$,

$$m^2 x = (m\iota_H)(m\pi_H)(x) + (m\iota_K)(m\pi_K)(x) \in H' \oplus K'.$$

Hence $n = m^2 G \subseteq H' + K' \subseteq G$. Suppose that $x \in H' \cap K'$. Then $x = (m\iota_H)(h) = (m\iota_K)(k)$ for some $h \in H$ and some $k \in K$. Using that $\pi_K\iota_H = 0$ and $\pi_H\iota_K = 0$ we obtain

$$m^2 x = (m\iota_H)(m\pi_H)((m\iota_K)(k)) + (m\iota_K)(m\pi_K)((m\iota_H)(h)) = 0.$$

Hence $H' \cap K' = 0$. Finally, the maps $H \ni x \mapsto (m\iota_H)(x) \in H'$ and $K \ni x \mapsto (m\iota_K)(x) \in K'$ are isomorphisms because they are surjective by definition and also injective because, e.g., if $h \in H$ and $m\iota_H(h) = 0$, then $0 = (m\pi_H)(m\iota_H)(h) = m^2 1_H(h) = m^2 h$, so $h = 0$. □

Definition 3.10. A decomposition $G \in H \oplus_{qu} K$ in $\mathbb{Q}\mathcal{A}$ is called a **quasi–decomposition**. A group H is called a **quasi–summand** of G if and only if there exists a quasi–decomposition $G \in H \oplus_{qu} K$. An indecomposable group of $\mathbb{Q}\mathcal{A}$ is called **strongly indecomposable**.

If $A \in A_1 \oplus A_2$ in a preadditive category $\mathcal{C}$ with structural maps $\iota_i : A_i \to A$, $\pi_i : A \to A_i$ satisfying $\pi_i \iota_i = 1_{A_i}$, $\pi_j \iota_i = 0_{A_i}$ for $i \neq j$, and $\iota_1 \pi_1 + \iota_2 \pi_2 = 1_A$, then $e_i := \iota_i \pi_i$ is an idempotent in $\mathrm{Hom}_{\mathcal{C}}(A, A)$.

Given an idempotent e in a category, it is not automatic that it can be factored as $e = \iota\pi$, $\pi\iota = 1$ in the fashion of the previous paragraph and that it produces a decomposition.

Definition 3.11. Let G be an object in a category $\mathcal{C}$ and $e = e^2 \in \mathrm{Hom}_{\mathcal{C}}(G, G)$. Then the **idempotent** e **splits in** $\mathcal{C}$ if there exists an object H and mappings $\iota \in \mathrm{Hom}_{\mathcal{C}}(H, G)$, $\pi \in \mathrm{Hom}_{\mathcal{C}}(G, H)$ such that $\iota\pi = e$ and $\pi\iota = 1_H$. We say that **idempotents split** in $\mathcal{C}$ if all idempotents in $\mathcal{C}$ are splitting.

We observe next that the splitting of idempotents means that every idempotent determines a direct decomposition.

Lemma 3.12. *Let $\mathcal{C}$ be a preadditive category. Suppose that A is an object of $\mathcal{C}$, and e is an idempotent in $\mathrm{Hom}_{\mathcal{C}}(A, A)$. Then $1_A - e$ is another idempotent of $\mathrm{Hom}_{\mathcal{C}}(A, A)$. Suppose further that there are objects and morphisms*

$$\iota_1 : A_1 \to A, \quad \pi_1 : A \to A_1, \quad \iota_2 : A_2 \to A, \quad \pi_2 : A \to A_2$$

such that

$$\iota_1\pi_1 = 1_{A_1}, \quad \pi_1\iota_1 = e, \quad \iota_2\pi_2 = 1_{A_2}, \quad \pi_2\iota_2 = 1_A - e.$$

Then $A \in A_1 \oplus A_2$ with structural maps $\iota_1, \pi_1, \iota_2, \pi_2$.

Proof. The claims are verified by straightforward computations that are left as an exercise. □

Lemma 3.13. *Idempotents split in $\mathbb{Q}\,\mathrm{End}(G)$.*

Proof. Let $G \in \mathbb{Q}\mathcal{A}$ and let $e = e^2 \in \mathbb{Q}\,\mathrm{End}(G)$. Write $e = \frac{1}{n} f$ with $f \in \mathrm{End}(G)$. It follows from $e = e^2$ that $nf = f^2$ and so $(n - f)f = 0$. We claim that

$$nG \subseteq (n - f)(G) \oplus f(G) \subseteq G.$$

In fact, let $x \in G$ be given. Then $nx = (n - f)(x) + f(x)$, hence $nG \subseteq (n - f)(G) + f(G) \subseteq G$. To show that the sum $(n - f)(G) + f(G)$ is direct, assume that $(n - f)(x) = f(y)$. Then $0 = f(f - n)(x) = f^2(y) = nf(y)$, and since G is torsion–free we obtain $(n - f)(x) = f(y) = 0$ as desired.

Now set $H := f(G)$, let $\iota_H : H \to G$ be the inclusion, let $p := \frac{1}{n}\iota_H \in \mathbb{Q}\,\mathrm{Hom}(H, G)$ and $q := f \in \mathbb{Q}\,\mathrm{Hom}(G, H)$. □

The ring $\mathbb{Q}\,\mathrm{End}(G)$ for $G \in \mathbb{Q}\mathcal{A}$ is a finite dimensional $\mathbb{Q}$–algebra, hence has finite length.

Proposition 3.14. *Let $G \in \mathbb{Q}\mathcal{A}$. Then G is indecomposable in $\mathbb{Q}\mathcal{A}$ if and only if $\mathbb{Q}\,\mathrm{End}(G)$ is local.*

Proof. (Arnold [2]) Let G be an indecomposable object in $\mathbb{Q}\mathcal{A}$. Then $\mathbb{Q}\operatorname{End}(G)$ contains no idempotents other than 0 and 1 since idempotents split in $\mathbb{Q}\mathcal{A}$. Also $\mathbb{Q}\operatorname{End}(G)$ is Artinian and therefore its radical $\operatorname{Rad}(\mathbb{Q}\operatorname{End}(G))$ is nilpotent. It follows that $\mathbb{Q}\operatorname{End}(G)/\operatorname{Rad}(\mathbb{Q}\operatorname{End}(G))$ contains no idempotents other than 0 and 1 since idempotents lift modulo the nilpotent radical $\operatorname{Rad}(\mathbb{Q}\operatorname{End}(G))$. Thus the quotient ring $\mathbb{Q}\operatorname{End}(G)/\operatorname{Rad}(\mathbb{Q}\operatorname{End}(G))$ is a semisimple Artinian ring without proper idempotents and therefore a division ring. Consequently, $\mathbb{Q}\operatorname{End}(G)$ is local. Conversely, a local ring contains no proper idempotents, hence G is indecomposable. □

The existence of indecomposable decompositions in $\mathbb{Q}\mathcal{A}$ is guaranteed by rank arguments. The uniqueness of decompositions is a consequence of the fact that strongly indecomposable groups have local quasi–endomorphism rings. So indecomposable quasi–decompositions are LE–decompositions of sorts. However, the arguments of module theory do not apply unchanged mainly because the category morphisms are not set–theoretic maps. The proper tool is provided by the so–called "Azumaya Unique Decomposition Theorem". A detailed proof is given in [25] for a preadditive category because this version was needed in the context of "almost completely decomposable groups". The Azumaya Theorem is usually proved for additive categories. The difference between additive and preadditive categories is the lack of a biproduct for every given finite set of objects in the latter but it is intuitively clear that the uniqueness question that deals with two given biproducts should not need the existence of arbitrary biproducts, only the existence of subsums.

Theorem 3.15 (Azumaya Unique Decomposition Theorem). *Let $\mathcal{C}$ be a preadditive category in which idempotents split. Suppose that $A \in A_1 \oplus \cdots \oplus A_m$ and also $A \in B_1 \oplus \cdots \oplus B_n$.*

1) *If each $\operatorname{Hom}_{\mathcal{C}}(A_i, A_i)$ is a local ring, then $B_j \in B_{j1} \oplus \cdots \oplus B_{js}$ and each B_{jk} is isomorphic with one of the A_i.*

2) *If each of the B_j is indecomposable, then $n = m$ and, after relabeling if necessary, $B_i \cong A_i$ for $i \in \{1, \ldots, n\}$.*

Corollary 3.16. *If $A \in A_1 \oplus \cdots \oplus A_m$ and also $A \in B_1 \oplus \cdots \oplus B_n$ are indecomposable decompositions in $\mathbb{Q}\mathcal{A}$, then $m = n$ and, after relabeling if necessary, $A_i \cong_{\mathrm{qu}} B_i$ for every i.*

By Corollary 3.2 we have the following result.

Theorem 3.17. *Let $A \in \mathbb{Q}\mathcal{A}$. Then* $\operatorname{Tot}(\mathbb{Q}\operatorname{End}(A)) = \operatorname{Rad}(\mathbb{Q}\operatorname{End}(A))$.

It is natural to investigate the relationship between $\operatorname{Tot}(\mathbb{Q}\operatorname{End}(A))$ and $\mathbb{Q}\operatorname{Tot}(\operatorname{End}(A))$ by which we mean the $\mathbb{Q}$–subspace of $\operatorname{End}(\mathbb{Q}A)$ generated by $\operatorname{Tot}(\operatorname{End}(A))$. Some connections are easily found.

Proposition 3.18. *Let $A \in \mathcal{A}$. Then the following statements are true.*

1) *If f is partially invertible in* $\mathrm{End}(A)$, *then f is partially invertible in* $\mathbb{Q}\,\mathrm{End}(A)$.
2) $\mathrm{End}(A) \cap \mathrm{Tot}(\mathbb{Q}\,\mathrm{End}(A)) \subseteq \mathrm{Tot}(\mathrm{End}(A))$.
3) $\mathrm{Tot}(\mathbb{Q}\,\mathrm{End}(A)) \subseteq \mathbb{Q}\,\mathrm{Tot}(\mathrm{End}(A))$.

Proof. 1) This is trivial since $\mathrm{End}(A) \subseteq \mathbb{Q}\,\mathrm{End}(A)$.

2) This is a formulation of the contrapositive of 1).

3) Let $f' \in \mathrm{Tot}(\mathbb{Q}\,\mathrm{End}(A))$. Then there is a positive integer n such that $f := nf' \in \mathrm{End}(A)$. We wish to show that $f \in \mathrm{Tot}(\mathrm{End}(A))$. Suppose by way of contradiction that f is pi in $\mathrm{End}(A)$. Then we have $0 \neq fg = e = e^2$ in $\mathrm{End}(A)$ and hence $f'(ng) = e = e^2$ in $\mathbb{Q}\,\mathrm{End}(A)$ saying contrary to assumption that f' is pi. □

Every idempotent in $\mathrm{End}(A)$ is an idempotent in $\mathbb{Q}\,\mathrm{End}(A)$ but in addition every $e \in \mathrm{End}(A)$ with $ne = me^2$ for non–zero integers m, n, or equivalently, $\frac{m}{n}e = \left(\frac{m}{n}e\right)^2$ produces an idempotent in $\mathbb{Q}\,\mathrm{End}(A)$. So presumably, there are pi elements in $\mathbb{Q}\,\mathrm{End}(A)$ on top of those of $\mathrm{End}(A)$, and accordingly, $\mathrm{Tot}(\mathbb{Q}\,\mathrm{End}(A))$ should be a proper subset of $\mathbb{Q}\,\mathrm{Tot}(\mathrm{End}(A))$.

We confirm these hunches with an example.

Example 3.19.

- Let V be a $\mathbb{Q}$–vector space with basis $\{v_1, v_2\}$, so $V = \mathbb{Q}v_1 \oplus \mathbb{Q}v_2$,
- let $A := \mathbb{Z}[3^{-1}]v_1 \oplus \mathbb{Z}[5^{-1}]v_2 \subseteq V$,
- and let $X := A + \mathbb{Z}\frac{1}{2}(v_1 + v_2) \subseteq V$.
- Set $\sigma := \mathrm{tp}(\mathbb{Z}[3^{-1}])$ and $\tau := \mathrm{tp}(\mathbb{Z}[5^{-1}])$.

Then the following statements hold.

1) $2X \subseteq A \subseteq X$ and $[X : A] = 2$.
2) $X(\sigma) = A(\sigma) = \mathbb{Z}[3^{-1}]v_1$ and $X(\tau) = A(\tau) = \mathbb{Z}[5^{-1}]v_2$, consequently $A = X(\sigma) \oplus X(\tau)$ is a fully invariant subgroup of X.
3) $\mathrm{End}(A) = \{(r, s) \mid r \in \mathbb{Z}[3^{-1}], s \in \mathbb{Z}[5^{-1}]\}$ where the action of (r, s) on $r_1v_1 + r_2v_2 \in A$ is given by
$$(r, s)(r_1v_1 + r_2v_2) = rr_1v_1 + sr_2v_2.$$
4) $\mathrm{End}(X) = \{(m+2x, m+2y) \mid m \in \{0, 1\}, x \in \mathbb{Z}[3^{-1}], y \in \mathbb{Z}[5^{-1}]\} \subseteq \mathrm{End}(A)$, the action being componentwise multiplication as for A.
5) $\mathrm{Aut}(X) = \{(\pm 3^m, \pm 5^n) \mid m, n \in \mathbb{Z}\}$.

6) The idempotents of $\operatorname{End}(X)$ are exactly the elements $(0,0)$ and $(1,1)$. In particular, X is directly indecomposable.

7) $\operatorname{Tot}(\operatorname{End}(X)) = \operatorname{End}(X) \setminus \operatorname{Aut}(X)$.

8) The idempotents in $\mathbb{Q}\operatorname{End}(X)$ are exactly the four maps $(0,0)$, $(1,0)$, $(0,1)$, $(1,1)$.

9) $\operatorname{Tot}(\mathbb{Q}\operatorname{End}(X))$ is a proper subset of $\mathbb{Q}\operatorname{Tot}(\operatorname{End}(A))$.

Proof. 1) Obviously $2X \subseteq A$ and $[X : A] \leq 2$. A simple computation shows that $\frac{1}{2}(v_1 + v_2) \notin A$, so $[X : A] = 2$.

2) By Lemma 1.21 $A(\sigma) = \mathbb{Z}[3^{-1}]v_1$ and $A(\tau) = \mathbb{Z}[5^{-1}]v_2$. Also by Lemma 1.19.5) $A(\sigma) = A \cap X(\sigma)$ and $X(\sigma) = A(\sigma)_*^X$, but it is easily checked that $A(\sigma)$ is pure in X, so $X(\sigma) = A(\sigma)$. Similarly, $X(\tau) = A(\tau)$.

3) Since A contains a basis of V, every endomorphism of A extends uniquely to a linear mapping on V. But since $A(\sigma) = \mathbb{Z}[3^{-1}]v_1$ and $A(\tau) = \mathbb{Z}[5^{-1}]v_2$ are fully invariant in A, any $f \in \operatorname{End}(A)$ has the property that $f(v_1) = rv_1$ for some $r \in \mathbb{Z}[3^{-1}]$ and $f(v_2) = sv_2$ for some $s \in \mathbb{Z}[5^{-1}]$ and hence can be described by the pair (r,s) acting as claimed. It is clear that every such pair defines an endomorphism of A.

4) Every endomorphism of X restricts to an endomorphism of A that extends uniquely to a linear mapping of V whose restriction to X is the initial endomorphism. The conclusion is that $\operatorname{End}(X)$ consists of all endomorphisms (r,s) of A with the property that $(r,s)(X) \subseteq X$. But the latter condition is equivalent to

$$(r,s)\left(\frac{1}{2}(v_1+v_2)\right) = \frac{1}{2}(rv_1 + sv_2) \in X.$$

Thus $(r,s) \in \operatorname{End}(X)$ if and only if there exist $m \in \mathbb{Z}$, $x \in \mathbb{Z}[3^{-1}]$ and $y \in \mathbb{Z}[5^{-1}]$ such that

$$\frac{1}{2}(rv_1 + sv_2) = xv_1 + yv_2 + m\frac{1}{2}(v_1+v_2).$$

By comparison of the coefficients of the basis elements we find that

$$r = m + 2x, \quad s = m + 2y.$$

It is easily checked that an endomorphism of A of the form $(m+2x, m+2y)$ is an endomorphism of X. The integer m may be taken to be either 0 or 1 depending on whether m is even or odd since any even integer can be combined with the summands $2x$ and $2y$.

5) An automorphism of X must induce an automorphism of A whose first components must be an automorphism of $\mathbb{Z}[3^{-1}]$, hence multiplication by $\pm$ a 3–power, and similarly the second component must be $\pm$ a 5–power. It remains to show that any pair $(3^a, 5^b)$ can be rewritten in the form of 4). Suppose $b \geq 0$.

Then $\pm 3^b = \pm 1 + \pm(3^b - 1)$ with $\pm(3^b - 1)$ even, and if $b < 0$, then $\pm 3^b = \pm 1 + \pm(1 - 3^{-b})/3^{-b}$ with $\pm(1 - 3^{-b})$ even. The same idea works for 5).

6) The proof is similar to the more complicated proof of 8).

7) Example 1.6.

8) Certainly $(0,0)$ and $(1,1)$ are idempotents in $\mathbb{Q}\,\mathrm{End}(X)$ and so are $(1,0) = \frac{1}{2}(2,0)$ and $(0,1) = \frac{1}{2}(0,2)$ since $(2,0),(0,2) \in \mathrm{End}(X)$ by 4). Idempotents in $\mathrm{End}(X)$ must restrict to idempotents of $\mathrm{End}(A)$ and it is easily seen that $\mathrm{End}(A)$ contains no idempotents other than the mentioned four.

9) We wish to find an element $f \in \mathbb{Q}\,\mathrm{Tot}(\mathrm{End}(X))$ that is not contained in $\mathrm{Tot}(\mathbb{Q}\,\mathrm{End}(X))$, i.e., we wish to find $f \in \mathbb{Q}(\mathrm{End}(X))$ that is not of the form $\frac{1}{n}(\pm 3^a, \pm 5^b)$ but is pi in $\mathbb{Q}\,\mathrm{End}(X)$. Consider $(2,3) = \frac{1}{2}(4,6) \in \mathbb{Q}\,\mathrm{End}(X)$. It is easy to see that $(2,3) = \frac{1}{n}(\pm 3^a, \pm 5^b)$ is not solvable while $(2,3) \cdot \frac{1}{12}(6,4) = (1,1)$ shows that $(2,3)$ is pi in $\mathbb{Q}\,\mathrm{End}(X)$. $\square$

Bibliography

[1] D. M. Arnold. *Finite Rank Torsion Free Abelian Groups and Rings*, volume 931 of *Lecture Notes in Mathematics.* Springer Verlag, 1982.

[2] D. M. Arnold. *Abelian Groups and Representations of Finite Partially Ordered Sets*, volume 2 of *CMS Books in Mathematics.* Springer Verlag, 2000.

[3] R. Baer. Abelian groups without elements of finite order. *Duke Math. J.*, 3:68–122, 1937.

[4] K.I. Beidar. On rings with zero total. *Beiträge Algebra und Geometrie*, 38:233–239, 1997.

[5] K.I. Beidar and F. Kasch. Toto–modules. *Algebra Berichte*, 76:1–19, 2000.

[6] K.I. Beidar and F. Kasch. Good conditions for the total. In *International Symposium on Ring Theory (Kyongju, 1999)*, pages 43–65. Trends in Math., Birkhäuser Verlag, 2001.

[7] K.I. Beidar and R. Wiegand. Radicals induced by the total of rings. *Beitr. Algebra und Geometrie*, 38:149–159, 1997.

[8] B. Brown and N.H. McCoy. The maximal regular ideal of a ring. *Proc. Am. Math. Soc.*, 1:165–171, 1950.

[9] B. Charles. Sous–groupes fonctoriels et topologies. In *Etudes sur les Groupes Abéliens, Colloque sur la Théorie des Groupes abélian tenu à l'Université de Montpellier en juin 1967*, pages 75–92. Dunod/Springer, 1968.

[10] A. L. S. Corner. A note on rank and direct decompositions of torsion-free abelian groups. *Proc. Cambridge Philos. Soc.*, 57:230–33, 1961.

[11] L. Fuchs. *Infinite Abelian Groups, Vol. I, II.* Academic Press, 1970 and 1973.

[12] M. Harada. *Applications of factor categories to completely indecomposable modules.* Publ. Dép. Math. Lyon, 1974.

[13] B. Jonsson. On direct decompositions of torsion–free abelian groups. *Math. Scand.*, 5:230–235, 1957.

[14] B. Jonsson. On direct decompositions of torsion–free abelian groups. *Math. Scand.*, 7:361–371, 1959.

[15] F. Kasch. *Modules and Rings.* A translation from the German "Moduln und Ringe". Academic Press, 1982.

[16] F. Kasch. *Moduln mit LE–Zerlegung und Harada–Moduln.* Lecture Notes. München, 1982.

[17] F. Kasch. Partiell invertierbare Homomorphismen und das Total. *Algebra Berichte*, 60:37–44, 1988.

[18] F. Kasch. The total in the category of modules. In *General Algebra.* Elsevier Science Publisher, 1988.

[19] F. Kasch. Modules with zero total. In *Icor 2000.* Innsbruck, 2000.

[20] F. Kasch. Locally injective modules and locally projective modules. *Rocky Mountain J. Math.*, 32(4), 2002.

[21] F. Kasch and W. Schneider. The total of modules and rings. *Algebra Berichte*, 69:37–44, 1992.

[22] F. Kasch and W. Schneider. Exchange properties and the total. In *Advances in Ring Theory.* Birkhäuser, Boston, 1997.

[23] G. Kolettis. Homogeneously decomposable modules. In *Studies on Abelian Groups*, pages 223–238, Paris, 1968. Dunod, Springer-Verlag.

[24] A. Mader. On the automorphism group and the endomorphism ring of abelian groups. *Ann. Univ. Sci. Budapest*, 8:3–12, 1965.

[25] A. Mader. *Almost Completely Decomposable Groups*, volume 13 of *Algebra, Logic and Applications.* Gordon and Breach Science Publishers, 2000.

[26] S.H. Mohamed and B. Müller. Continuous and discrete modules. *Lecture Notes*, 147, 1990.

[27] W. Schneider. Das Total von Moduln und Ringen. *Algebra Berichte*, 55, 1987.

[28] J.M. Zelmanowitz. On the endomorphism ring of a discrete module: A theorem of F. Kasch. In *Advances in Ring Theory*, pages 317–322. Birkhäuser, Boston, 1997.

[29] A. Zöllner. Lokal–direkte Summanden. *Algebra Berichte*, 51:64pp, 1984.

Index

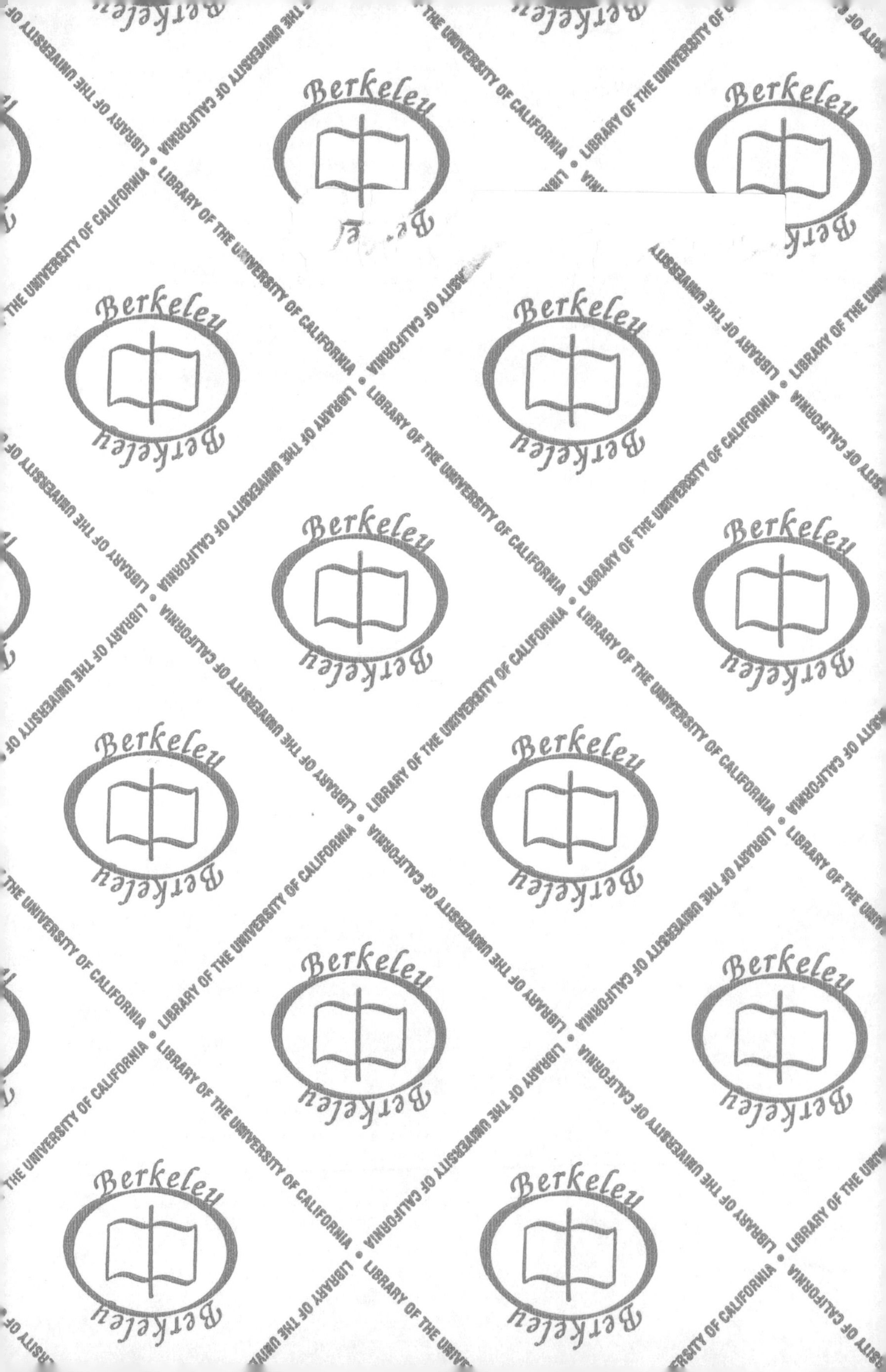
Berkeley
LIBRARY OF THE UNIVERSITY OF CALIFORNIA